Hi everyone, I wrote this document will help you understand more about SolidWorks, and use it more easily.

For example a Power Outlet, we will better understand commands like.

- ✓ Create 2d sketch.
- ✓ Extruded Bass
- ✓ Extruded Cut.
- ✓ Shell.
- ✓ Mirror.
- ✓ Linear Patemrn.
- ✓ Mate.
- ✓ Create Plane.

Prior to the examples, I should note a few things.

- ✓ Part is created by sketch.
- ✓ Sketch is the base to define của part, form and features.
- ✓ Before you start creating in sketches cần select the sketch plane or face where sẽ place on.
- ✓ After select the sketch plane or face Will Be, sketch on it!

Creat Part 1 with name Power Outlet-1

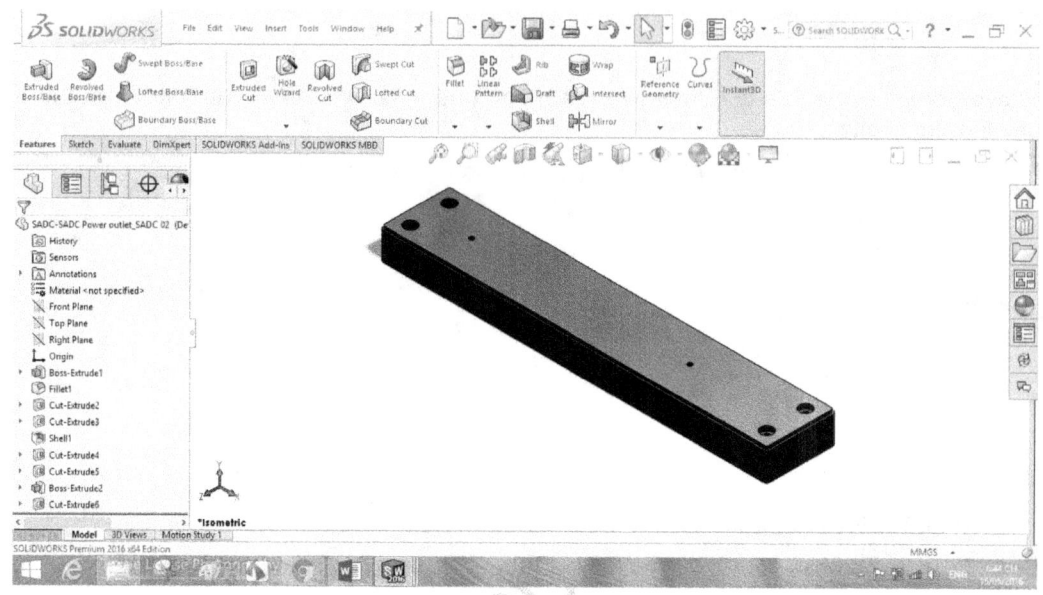

Step 1. Open Solidworks=>File=>New=>Part.

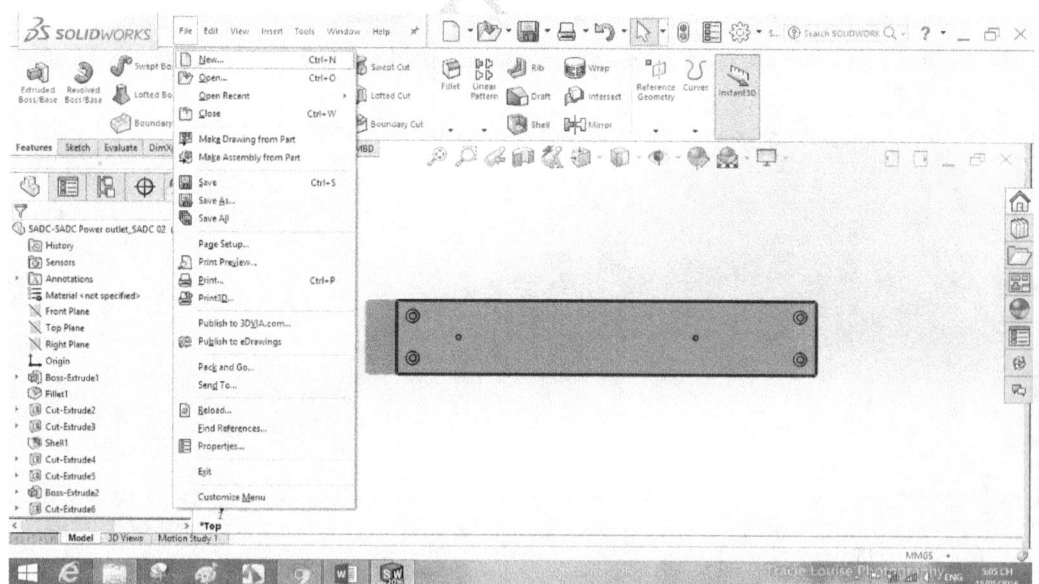

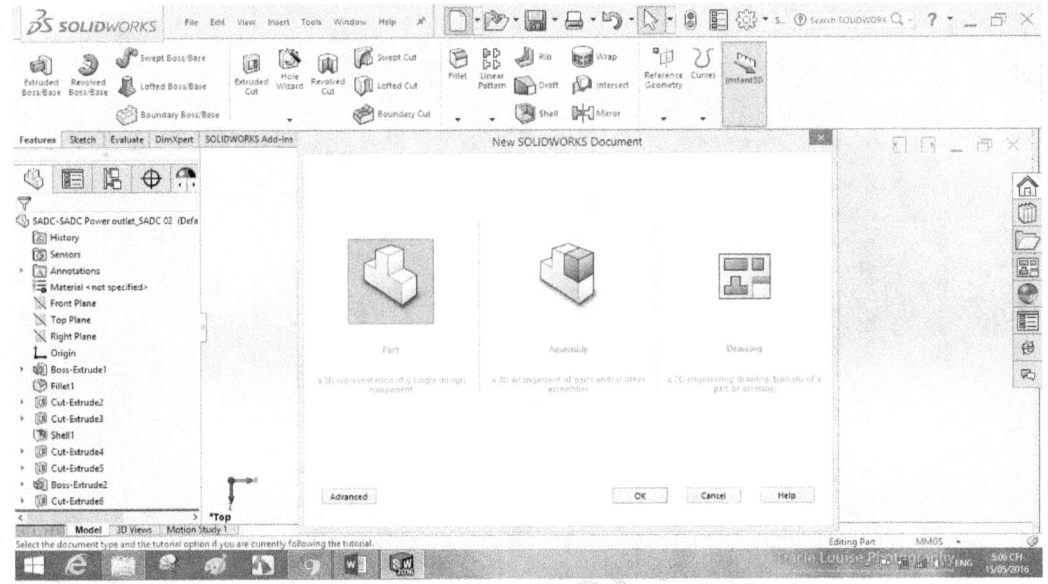

Step 2. Right Click Top Plane =>Sketch.

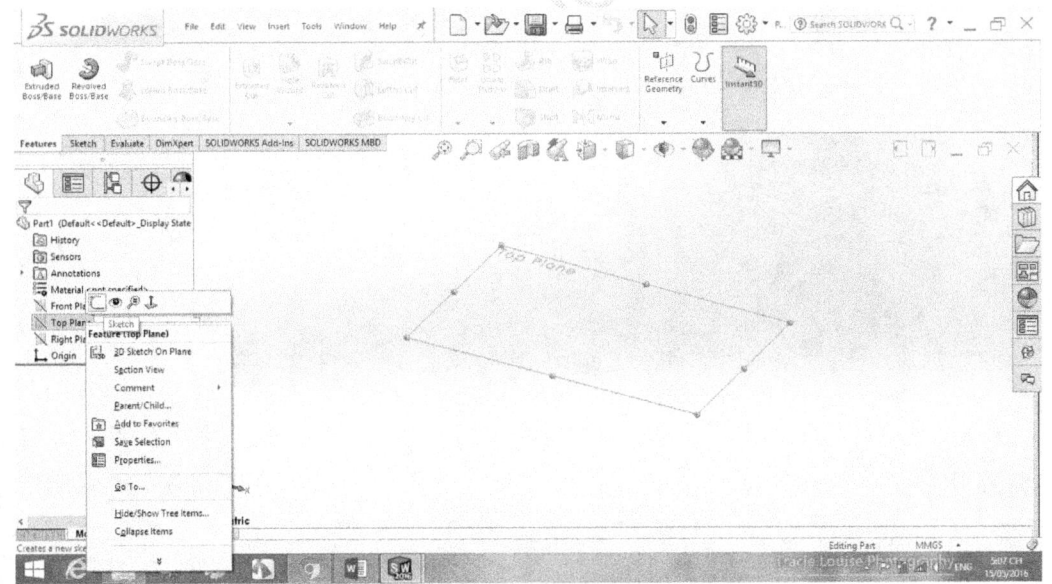

Step 3. Click Sketch=>Corner Rectangle.

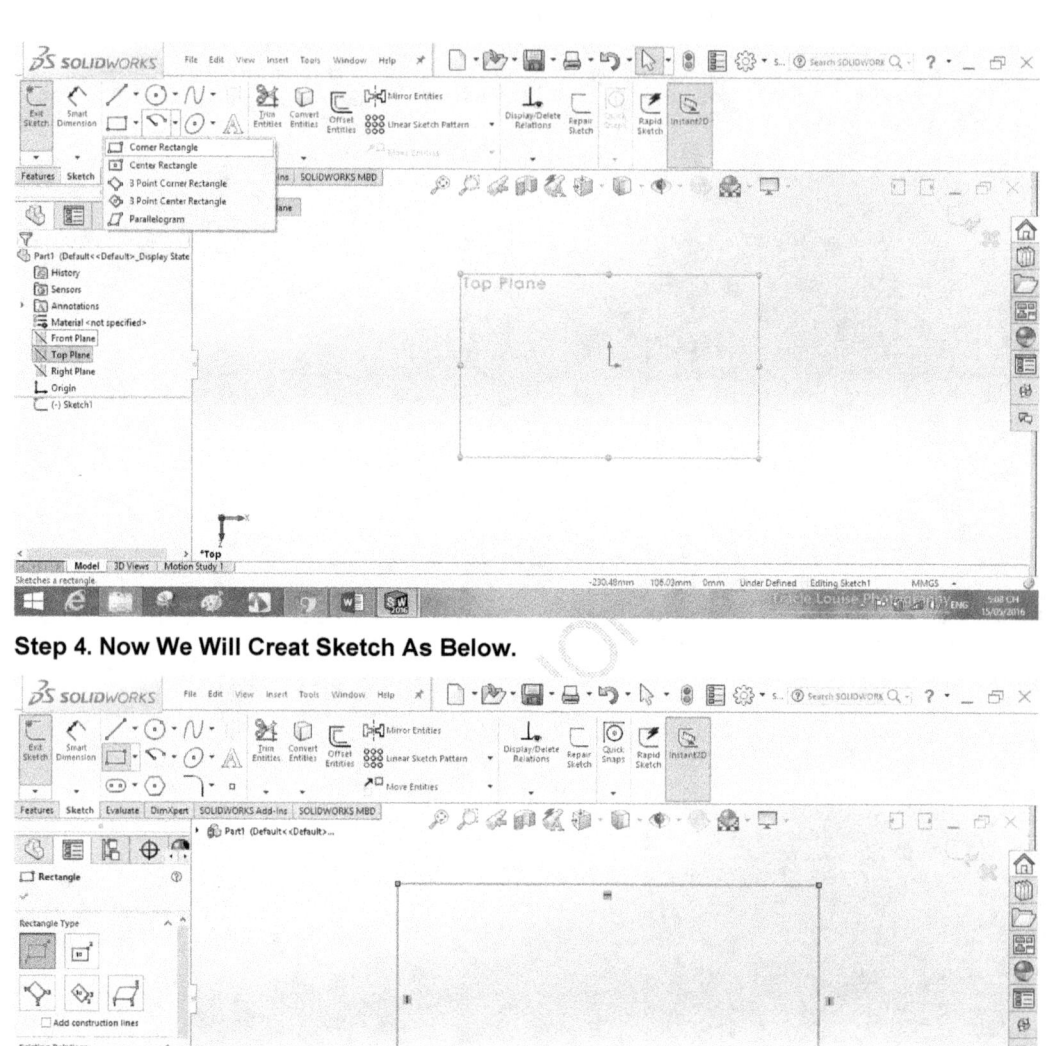

Step 4. Now We Will Creat Sketch As Below.

Step 5. Click Smart Dimension.

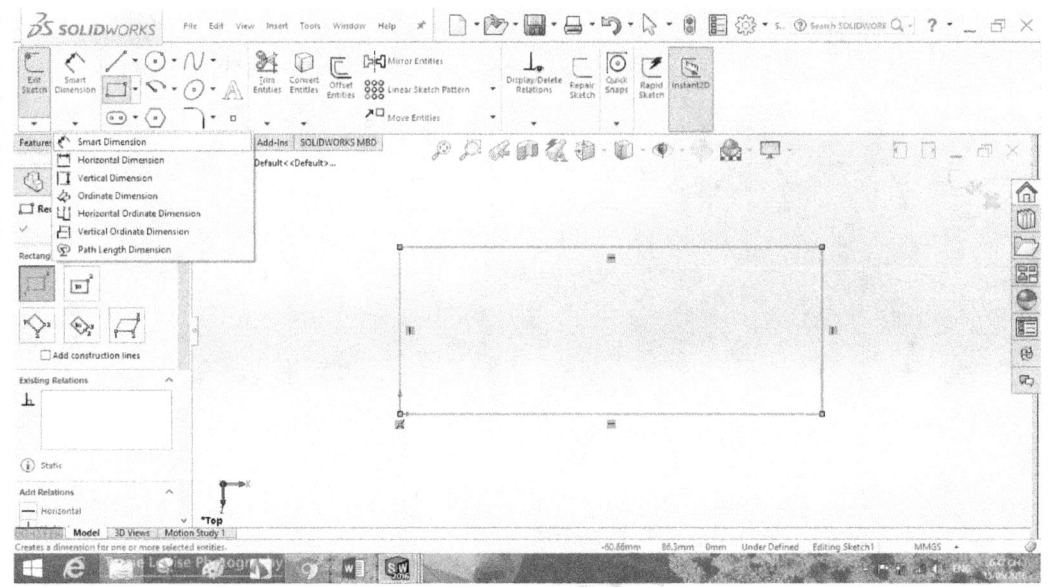

Step 6. And Constraints As Below Then Click

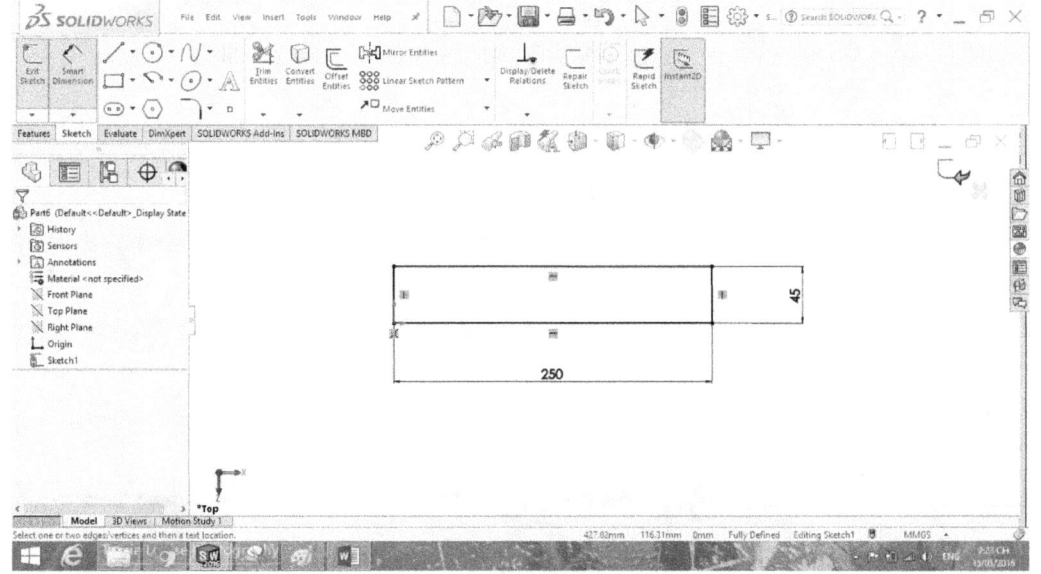

Step 7. Click Sketch Just Created=>Features=>Extruded Boss/Base.

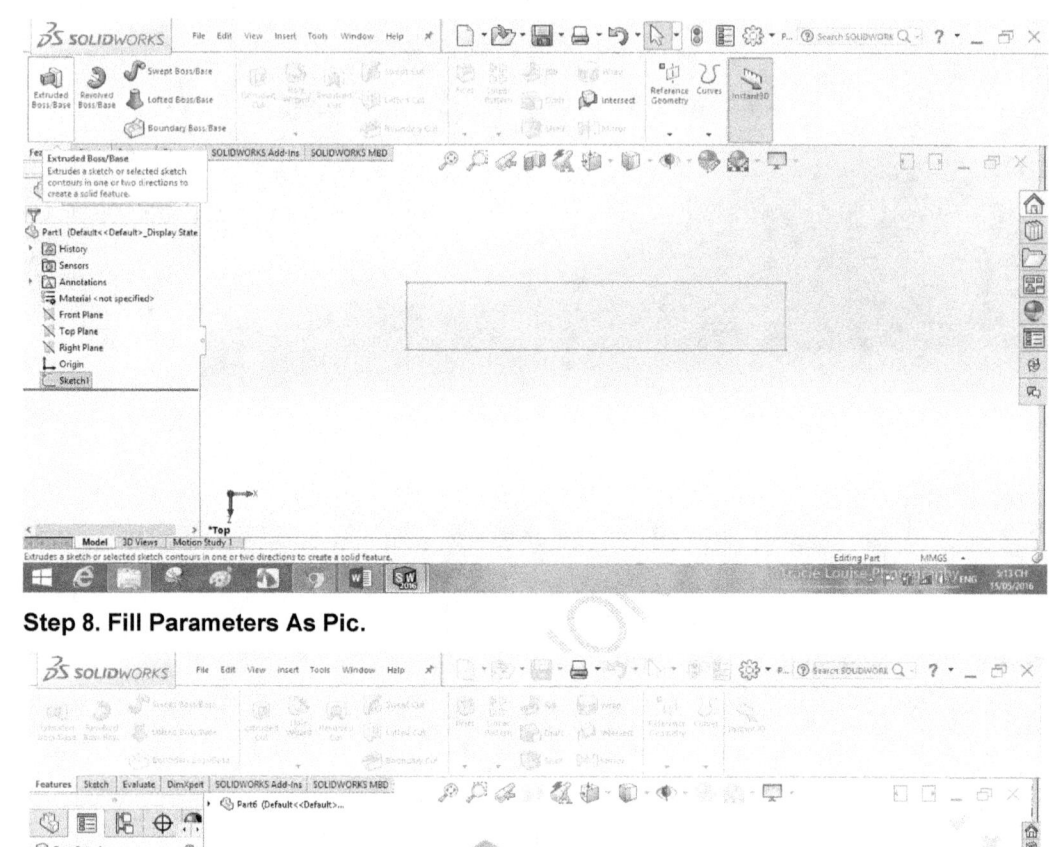

Step 8. Fill Parameters As Pic.

Step 9. Click Fillet To Fillet 4 Edge With R=2. Then Click Ok.

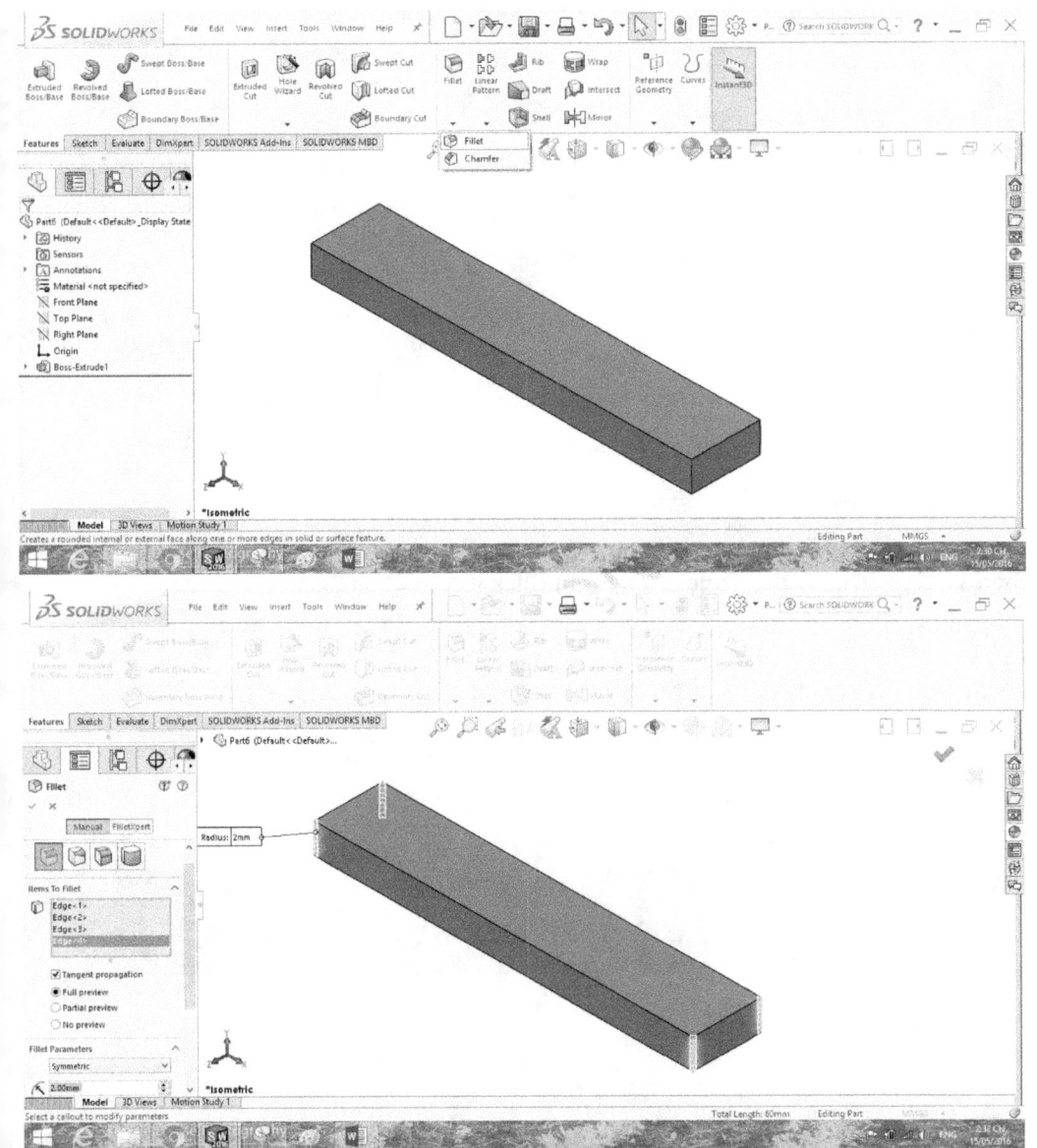

Step 10. Creat Sketch On Plane As Pic. (Ctrl + 8 To Creat Sketch)

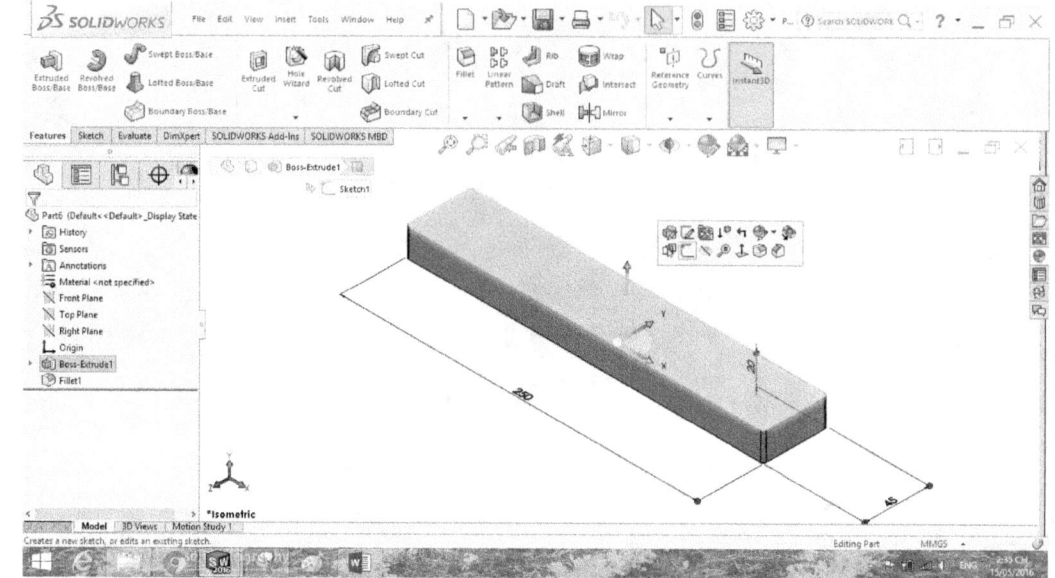

Step 11. Click One Outer Edge Then Right Click It And Click Select Tangency.

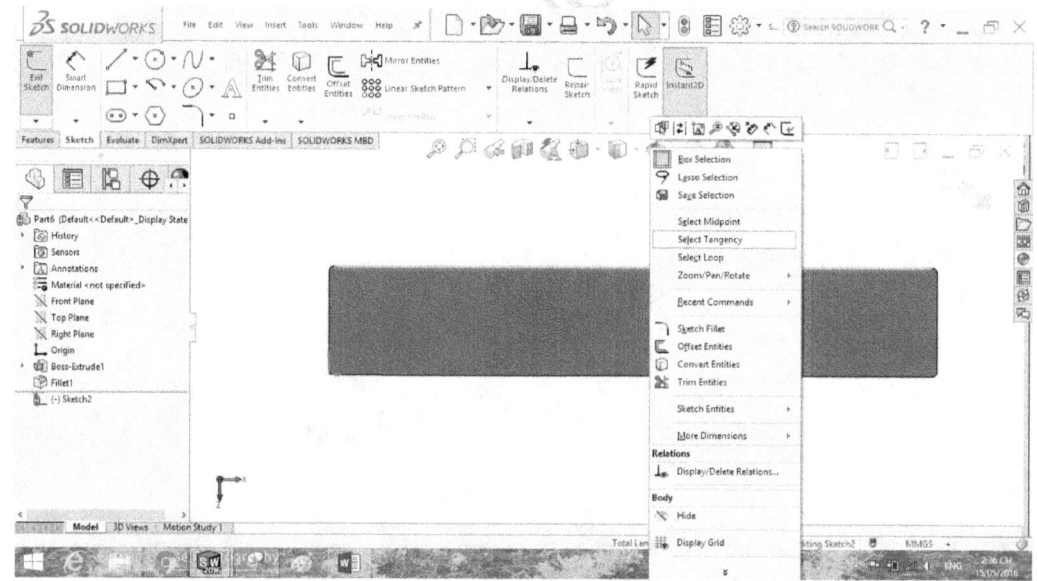

Step 12. Click Convert Entities.

Step 13. Continue, Click Sketch Just Created And Right Click It Then Click Select Chain.

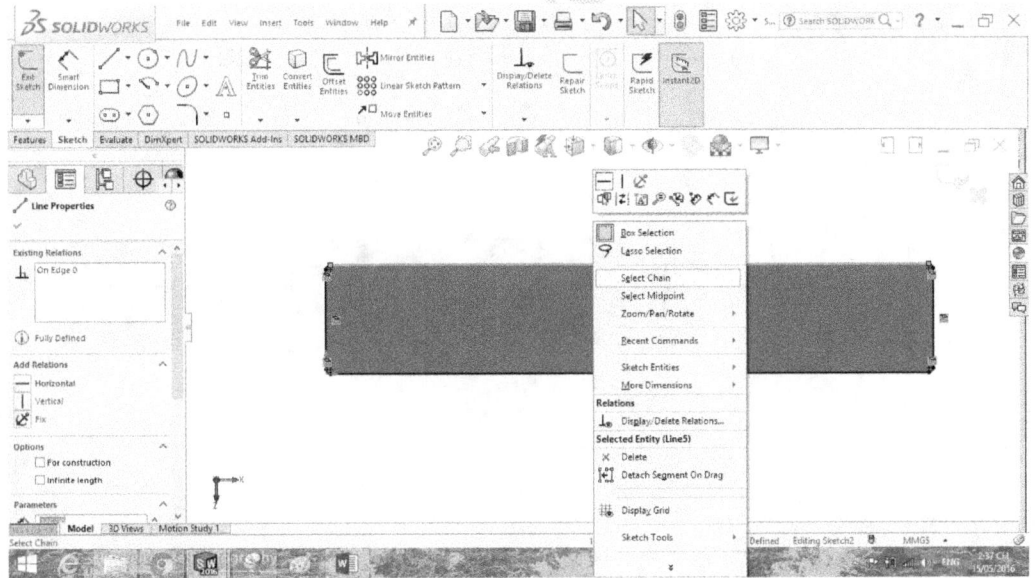

Step 14. Cilck Offset Entities. Fill D=1 And Tick Box As Pic. Then Click Ok.

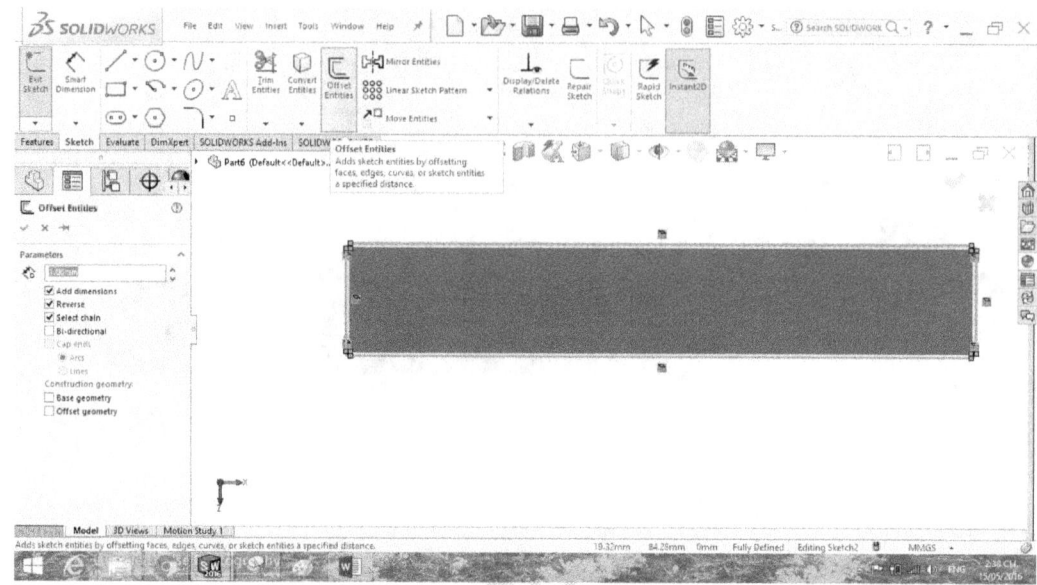

Step 15. Click sketch just created=>Features=>Extruded Cut.

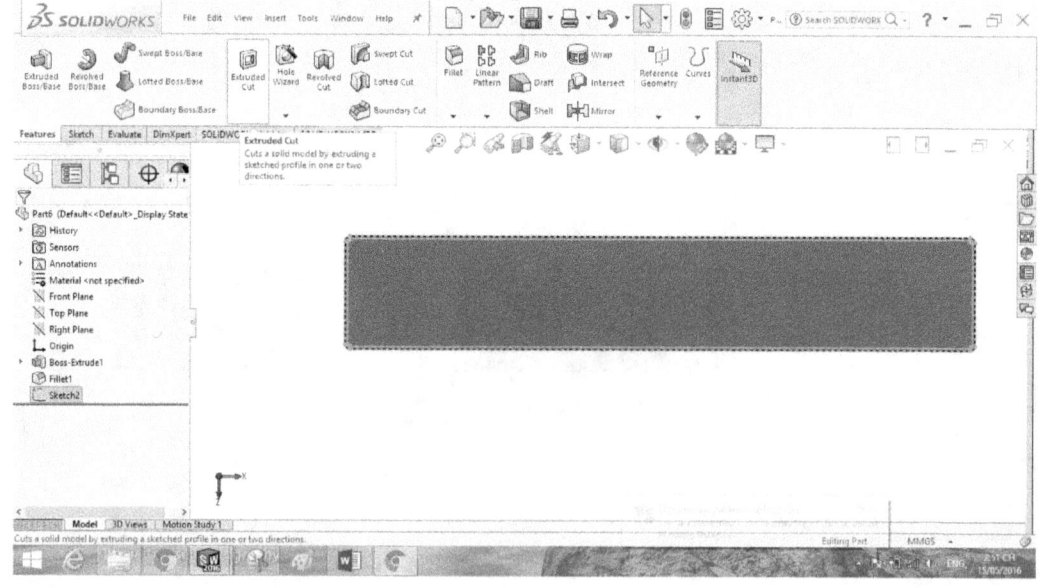

Step 16. Fill D1=2. Click Ok.

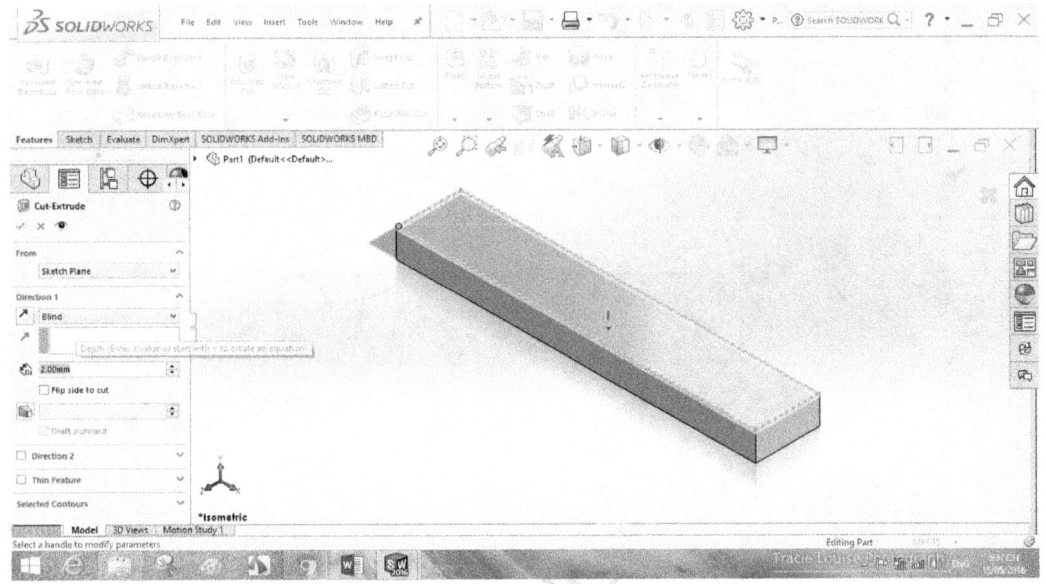

Step 17. Creat Sketch On Plane As Pic.

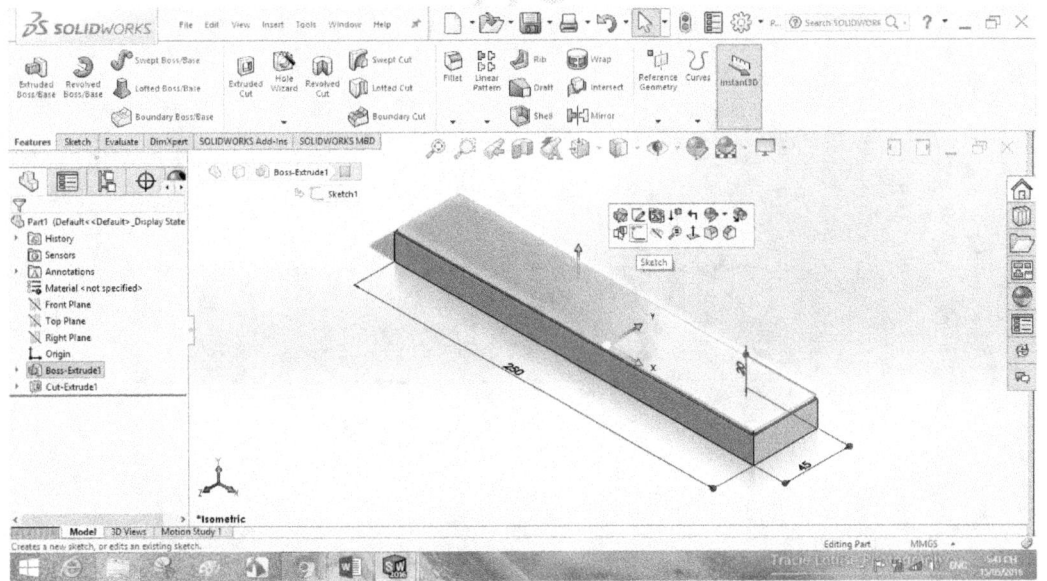

Step 18. Click Circle To Creat Sketch.

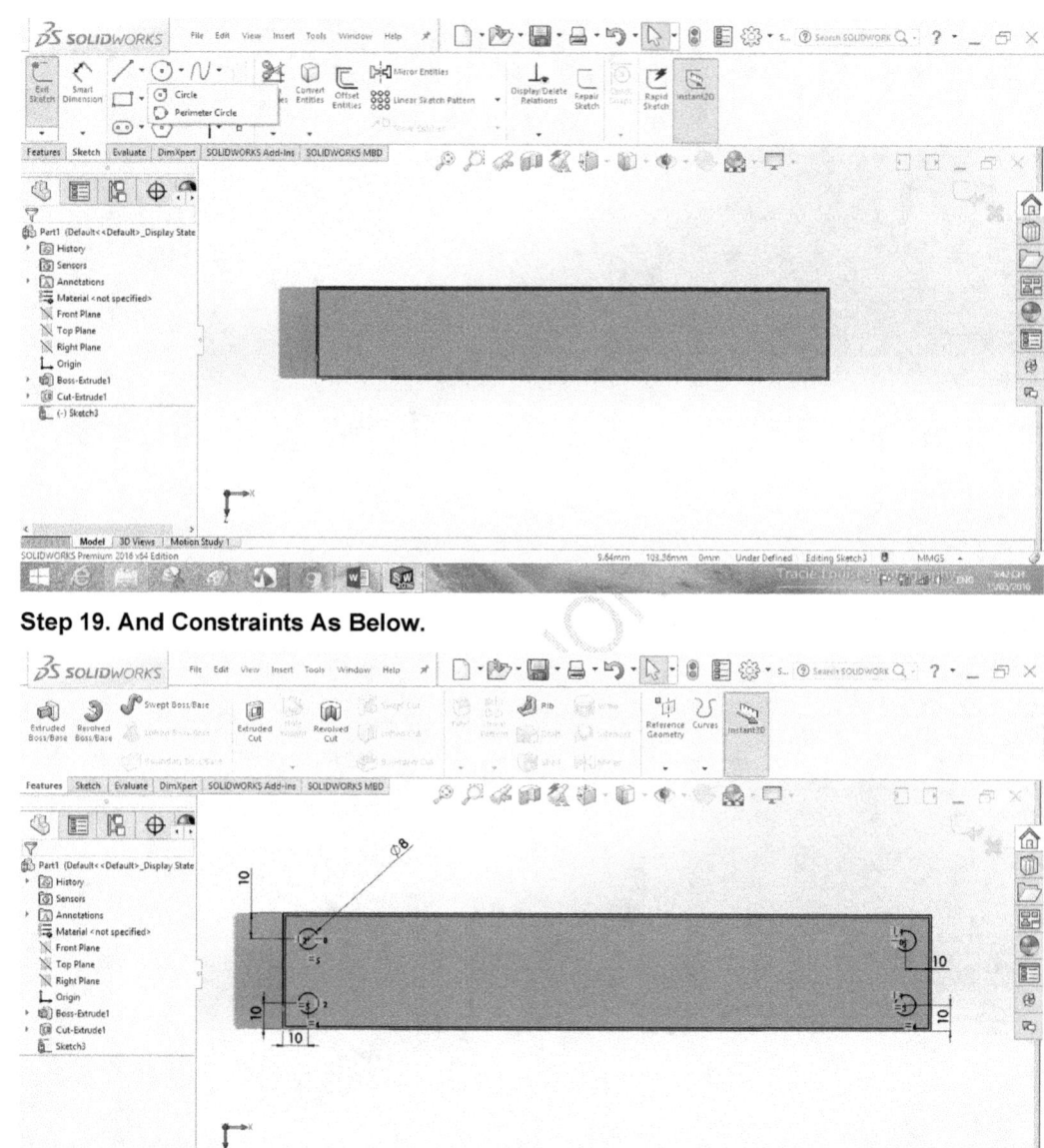

Step 19. And Constraints As Below.

Step 20. Click Sketch Just Created=>Features=>Extruded Cut. Fill D1=3.

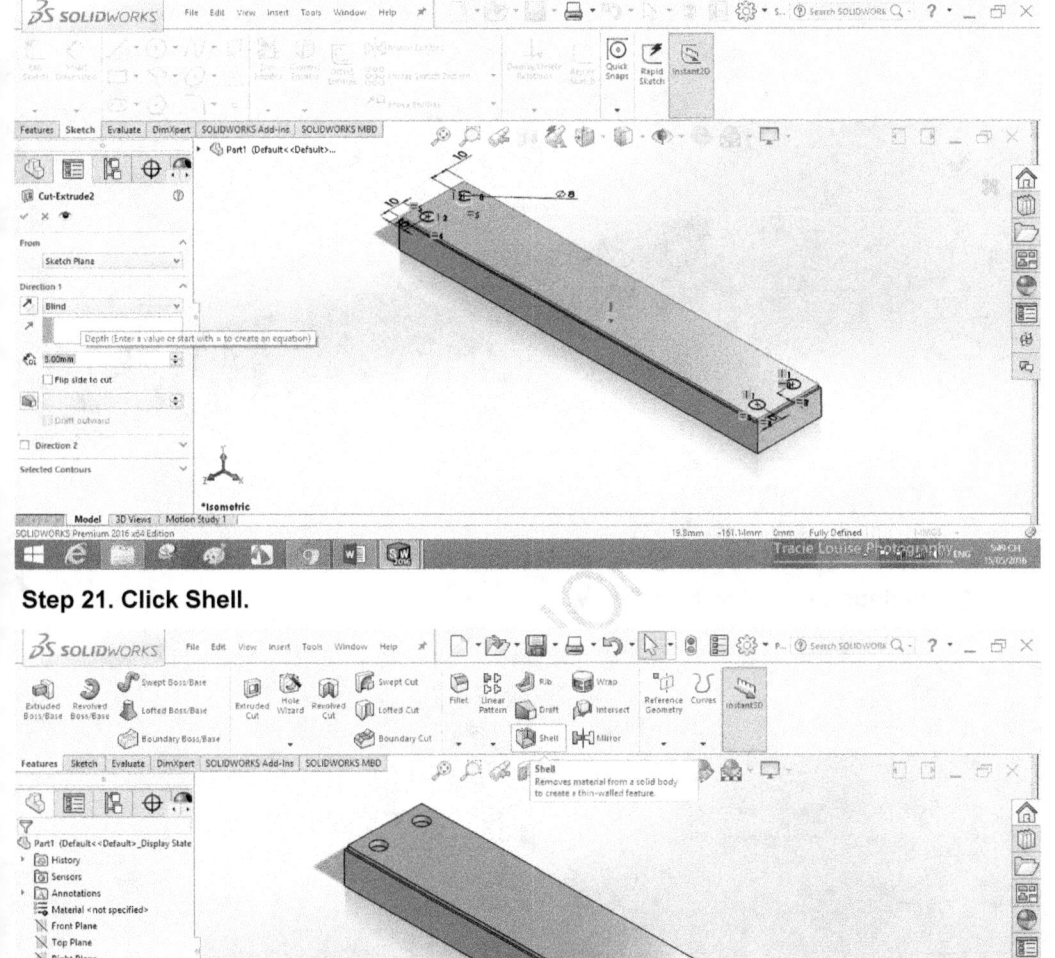

Step 21. Click Shell.

Step 22. Click As Pic And Fill D1=2.

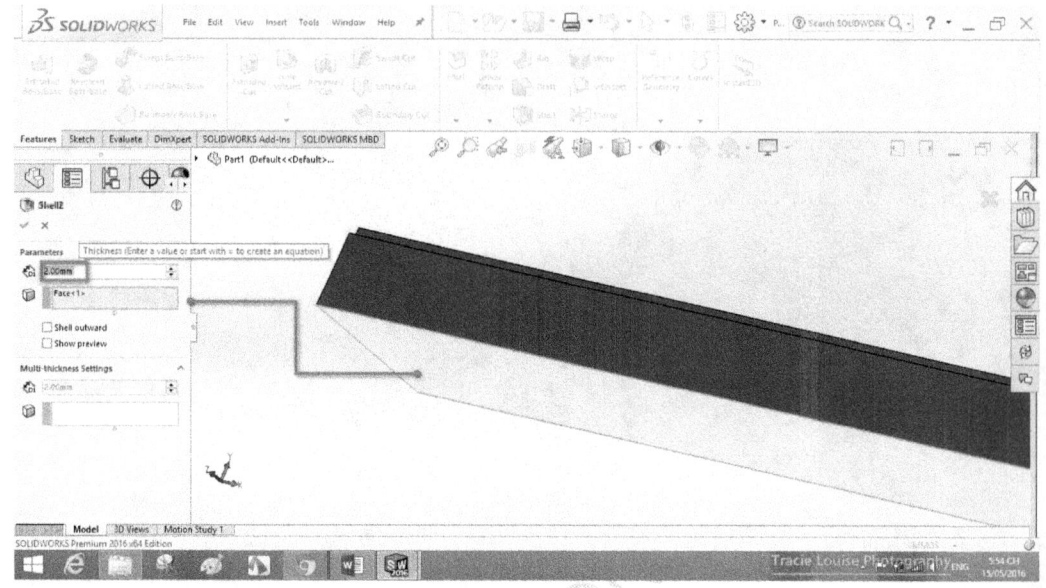

Step 23. Continue Creat Sketch As Below.

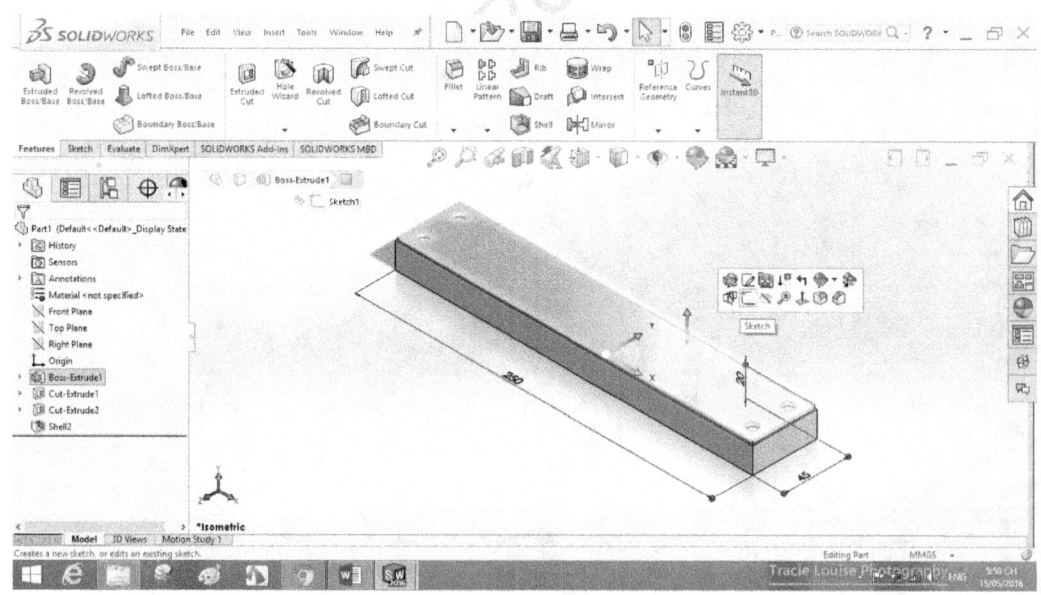

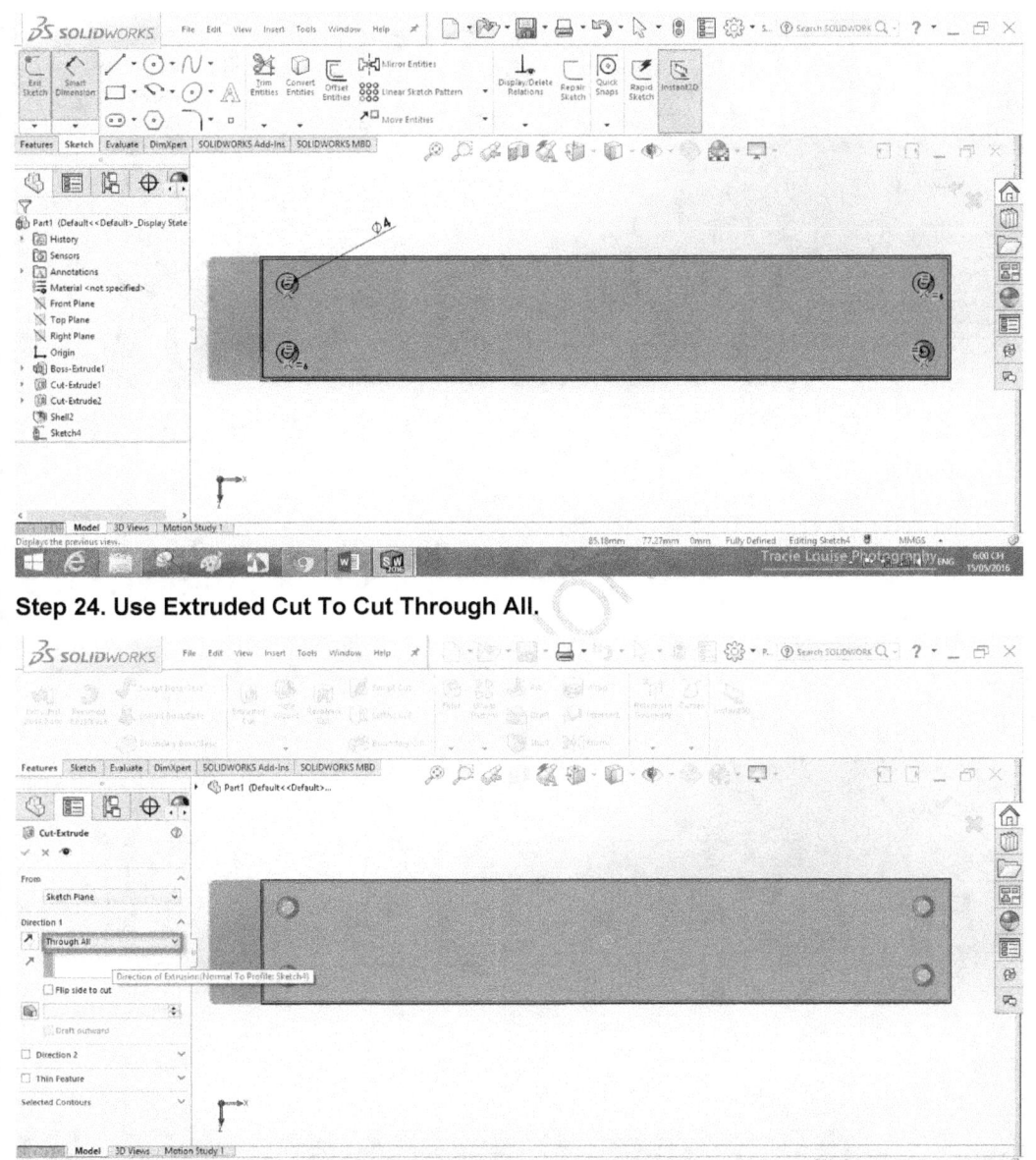

Step 24. Use Extruded Cut To Cut Through All.

Step 25. Continue Creat 1 Sketch And Use Extruded Cut To Cut Through All.

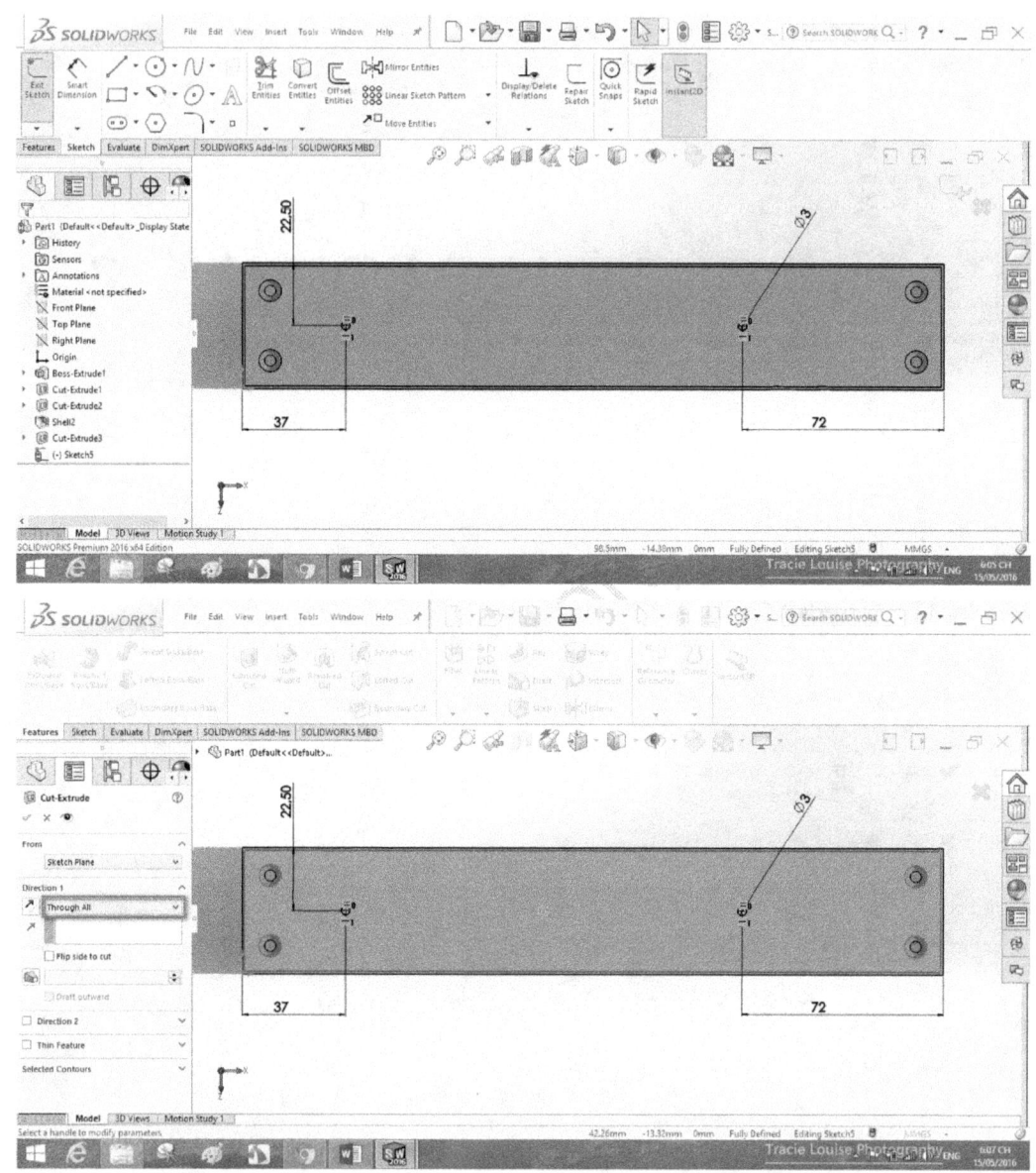

Step 26. Continue Creat More Sketch And Use Extruded Boss/Base with D1=10.

Note. Plane Is Pic.

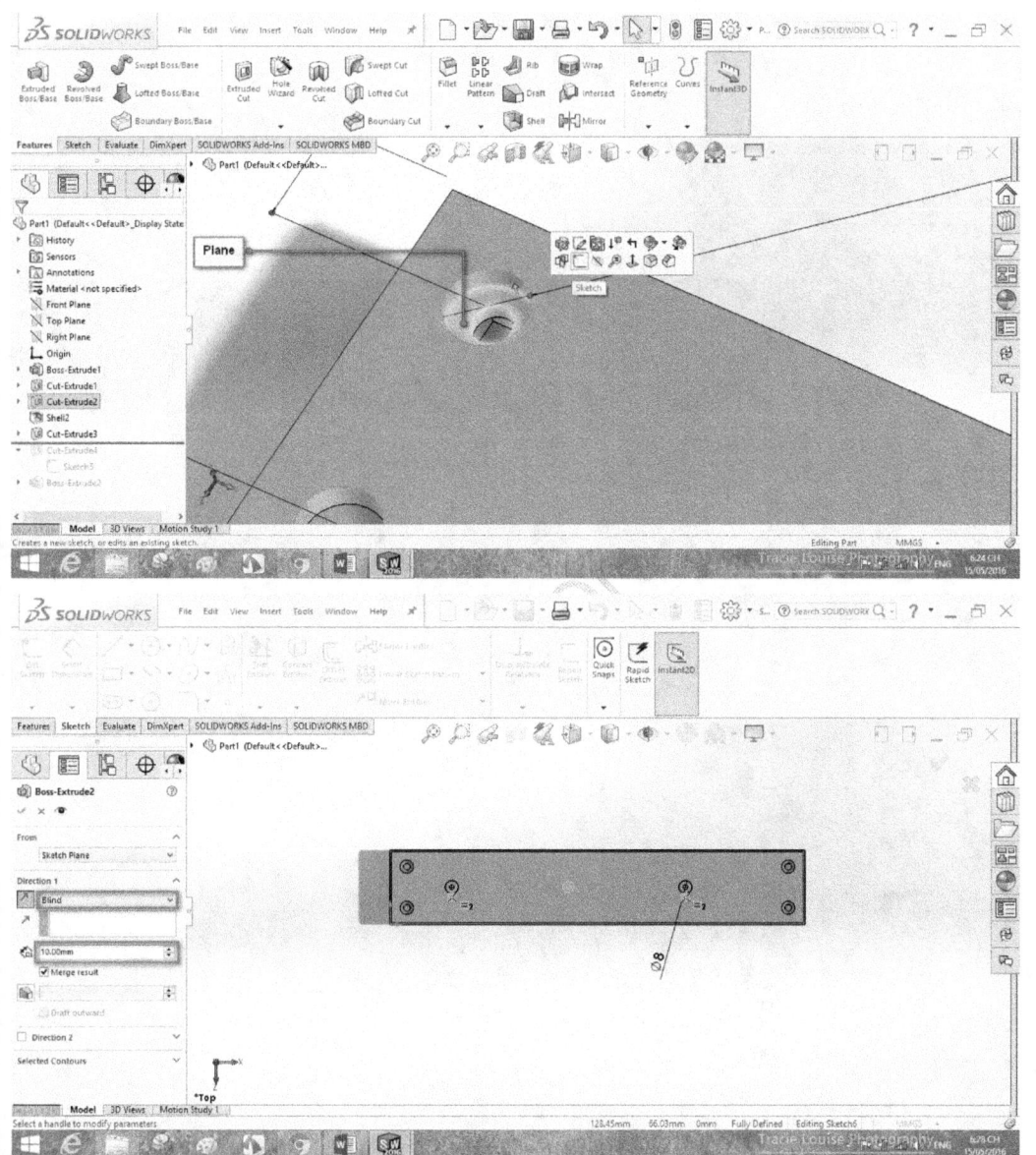

Step 27. Continue Creat Sketch And Use Extruded Cut To Cut Through All.

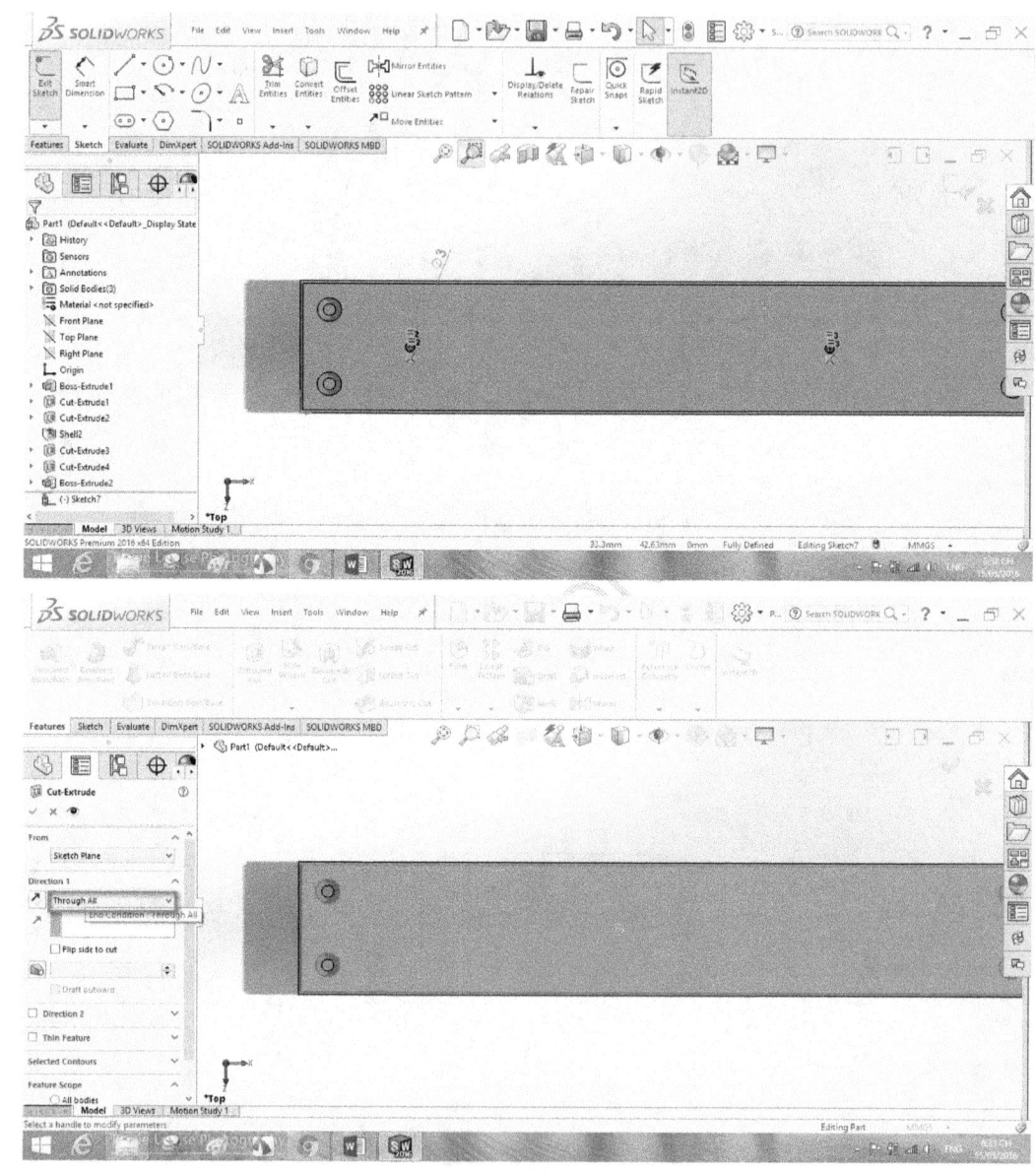

Step 28. Save As with Name Power Outlet-1.

Creat Part 2 with name Power Outlet-2

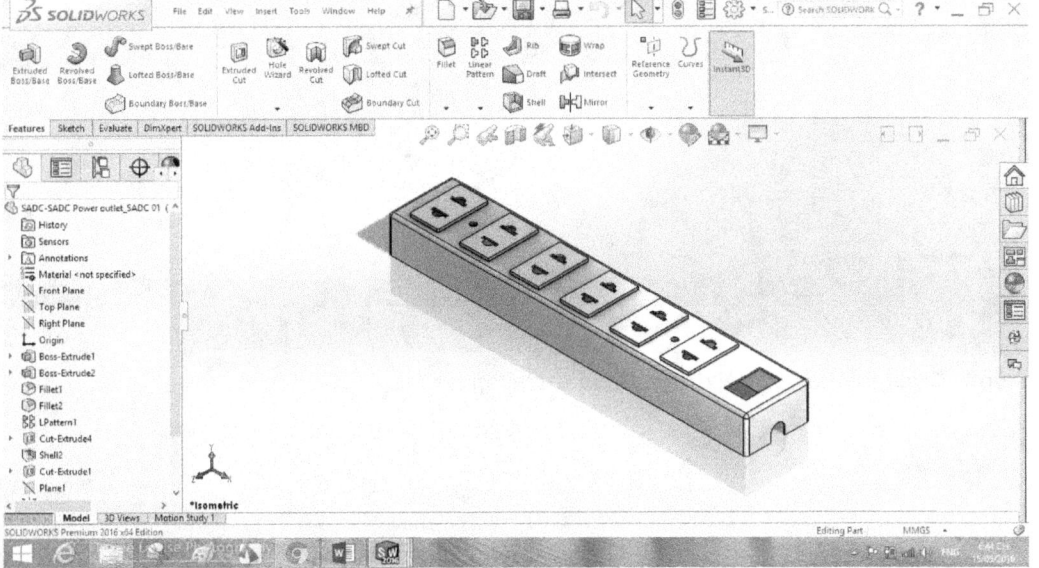

Step 1. Open Solidworks=>File=>New=>Part.

Step 2. Creat Sketch Rectangle (X=250, Y=45) On Top Plane, Use Extrude Boss with Depth=25.

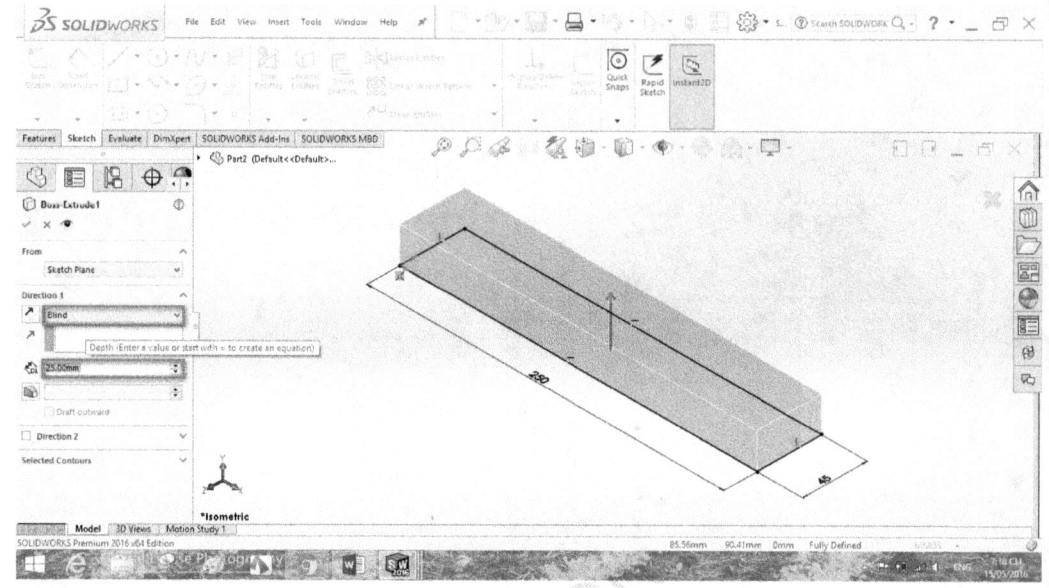

Step 3. Creat Sketch, Use Extrude Boss with Depth=2.

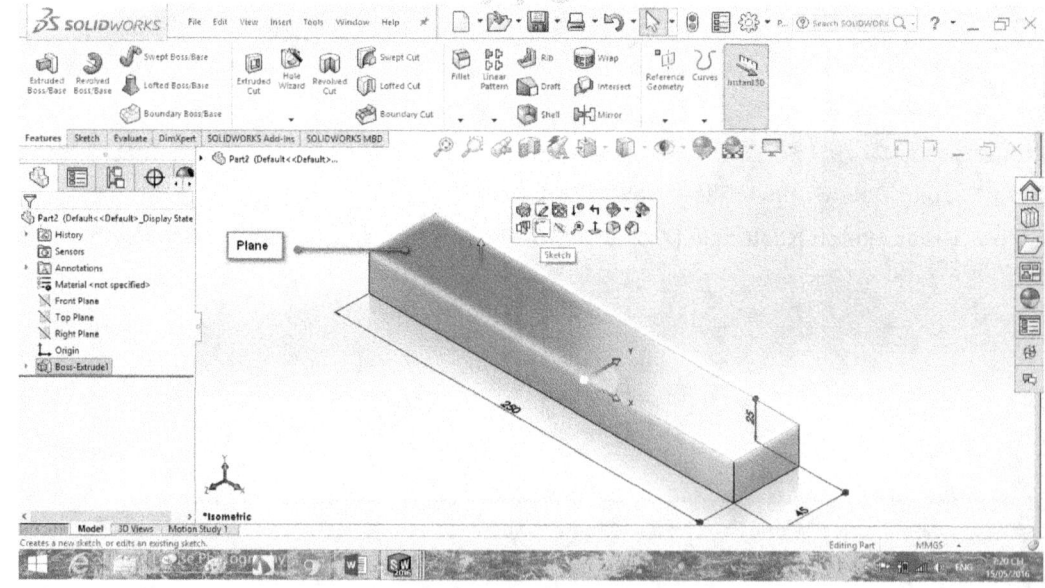

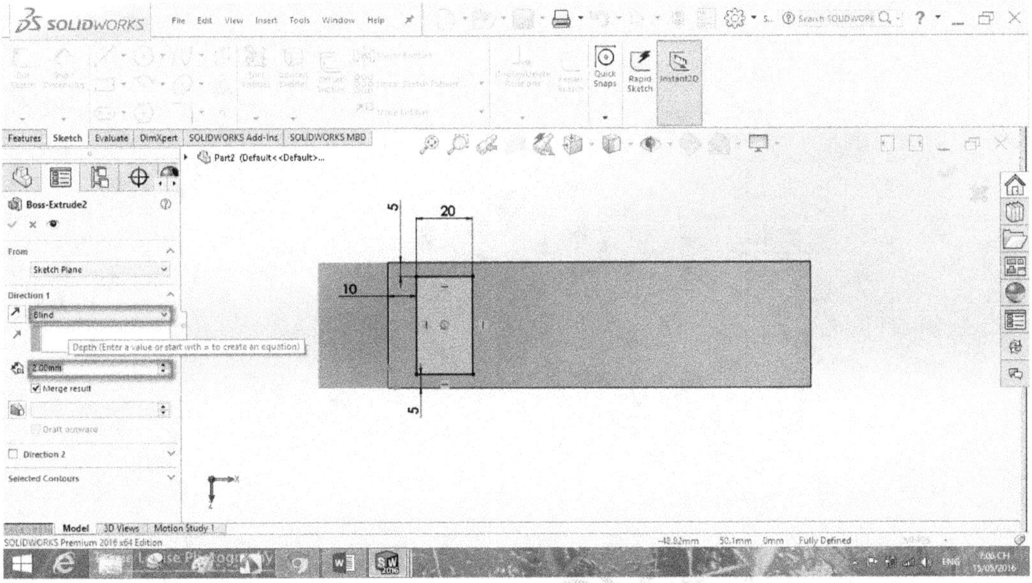

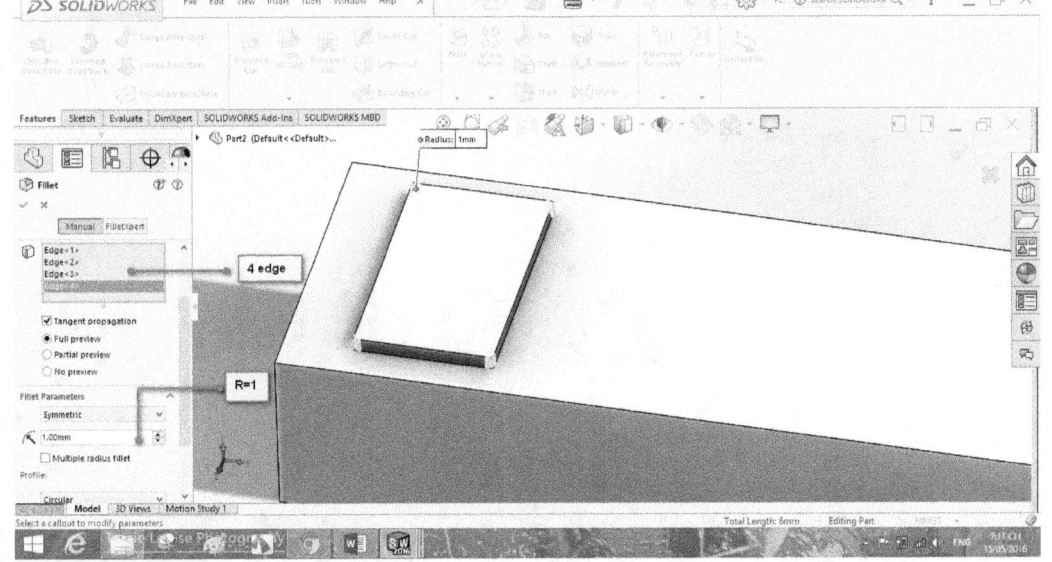

Step 4. Use 4 edge with R=1. See pic.

Step 5. Continue Fillet With R=2 As Below.

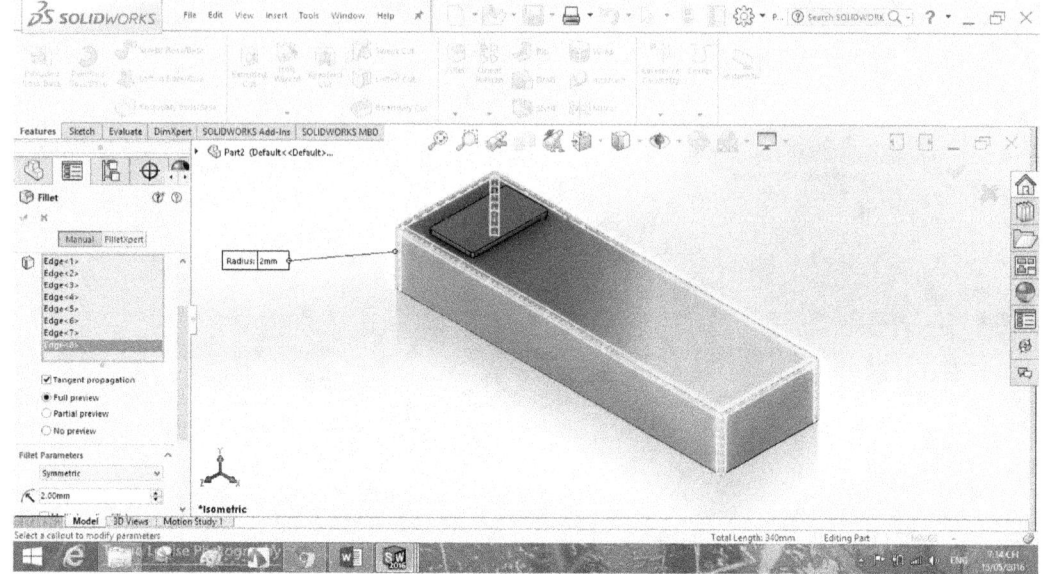

Step 6. Click Features=>Linear Pattern.

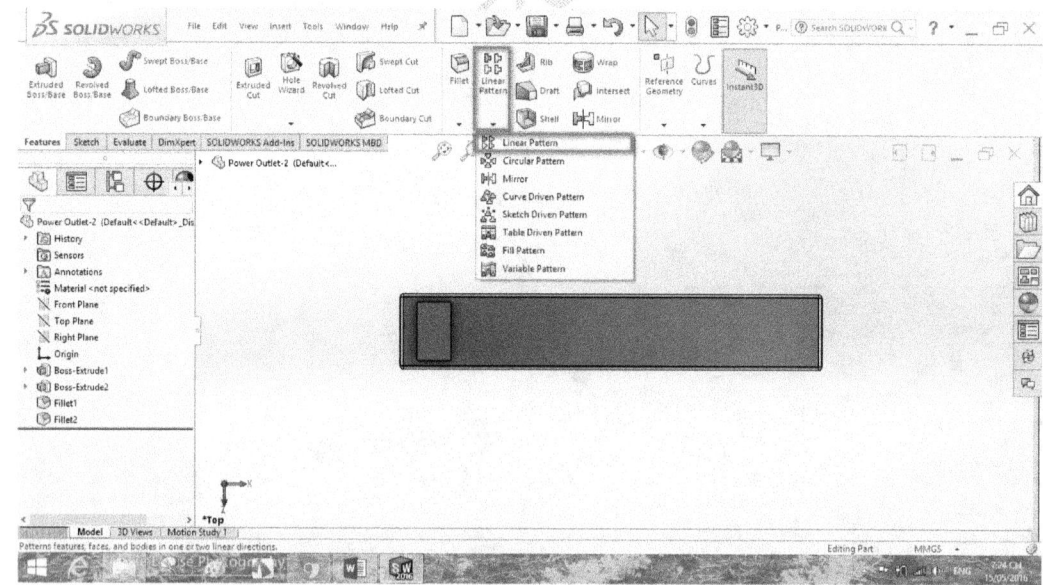

Step 7. Fill parameters as below.

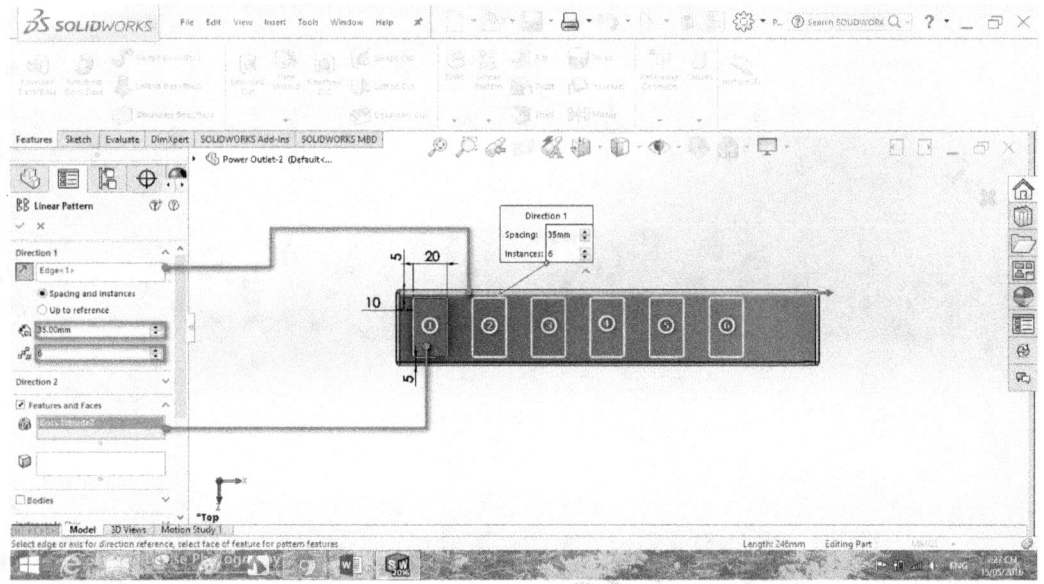

Step 8. Creat Sketch.

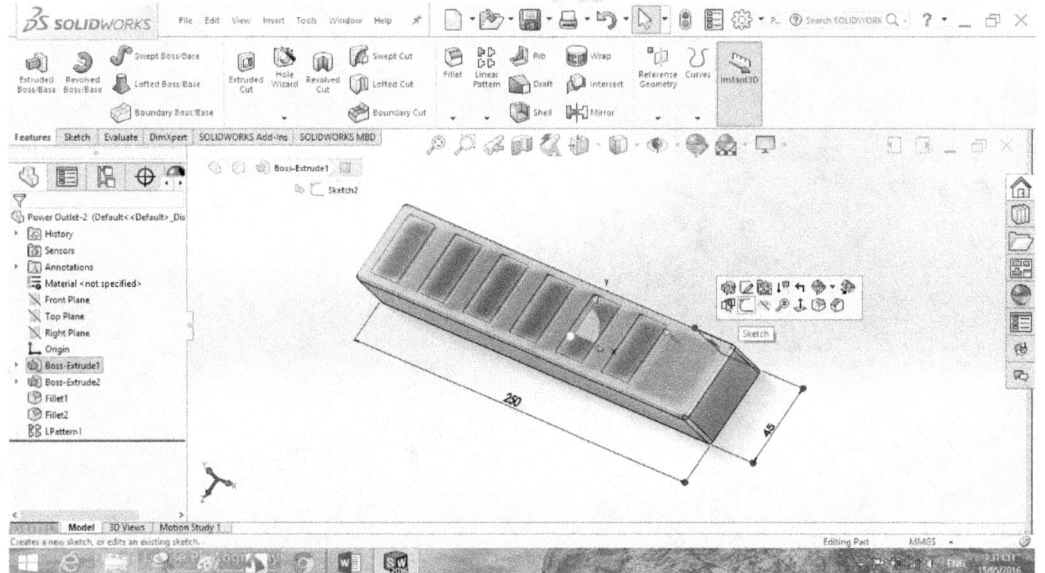

Step 8. Use Extruded/Boss and fill as is.

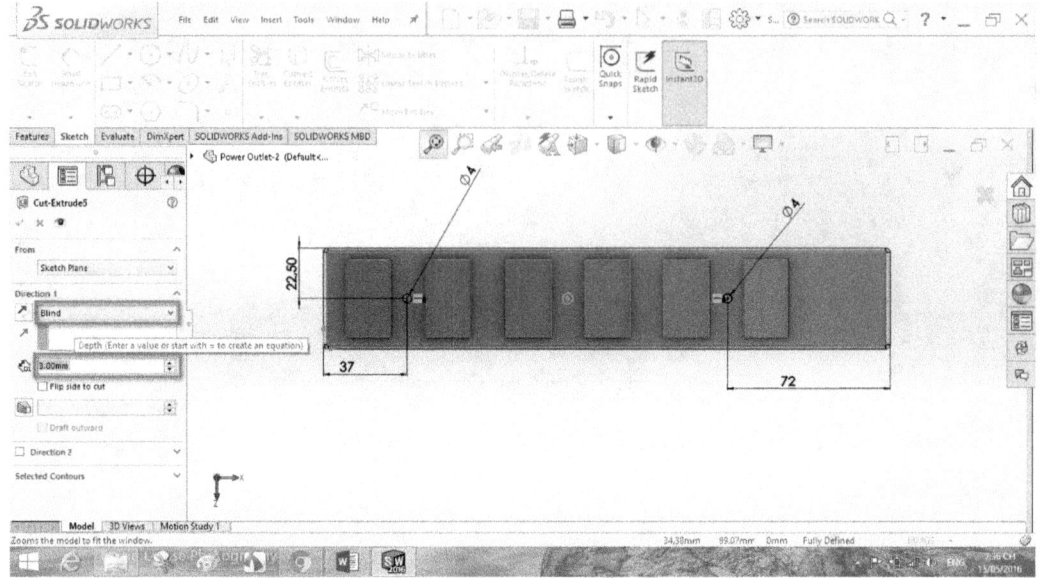

Step 9. Use Shell and fill as is.

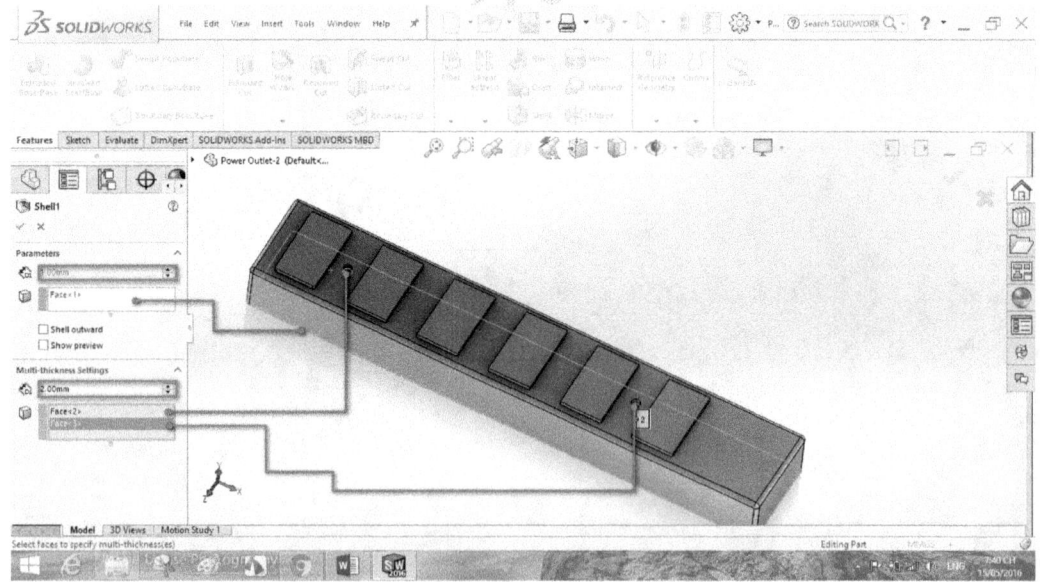

Step 10. Creat Sketch. Use Extruded Cut. Fill as is.

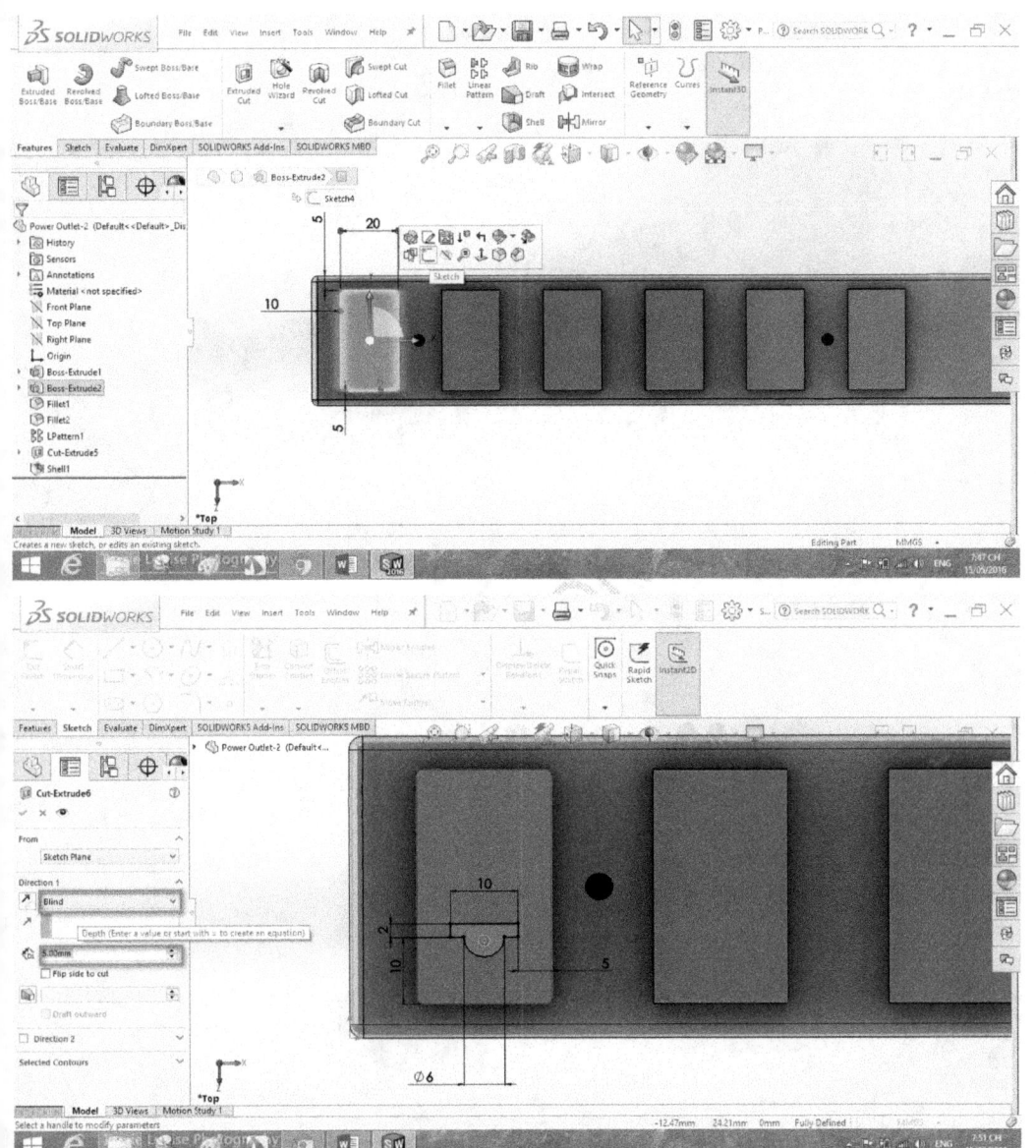

Step 11. Creat Plane. (Features=>Plane). Then Select Plane as below.

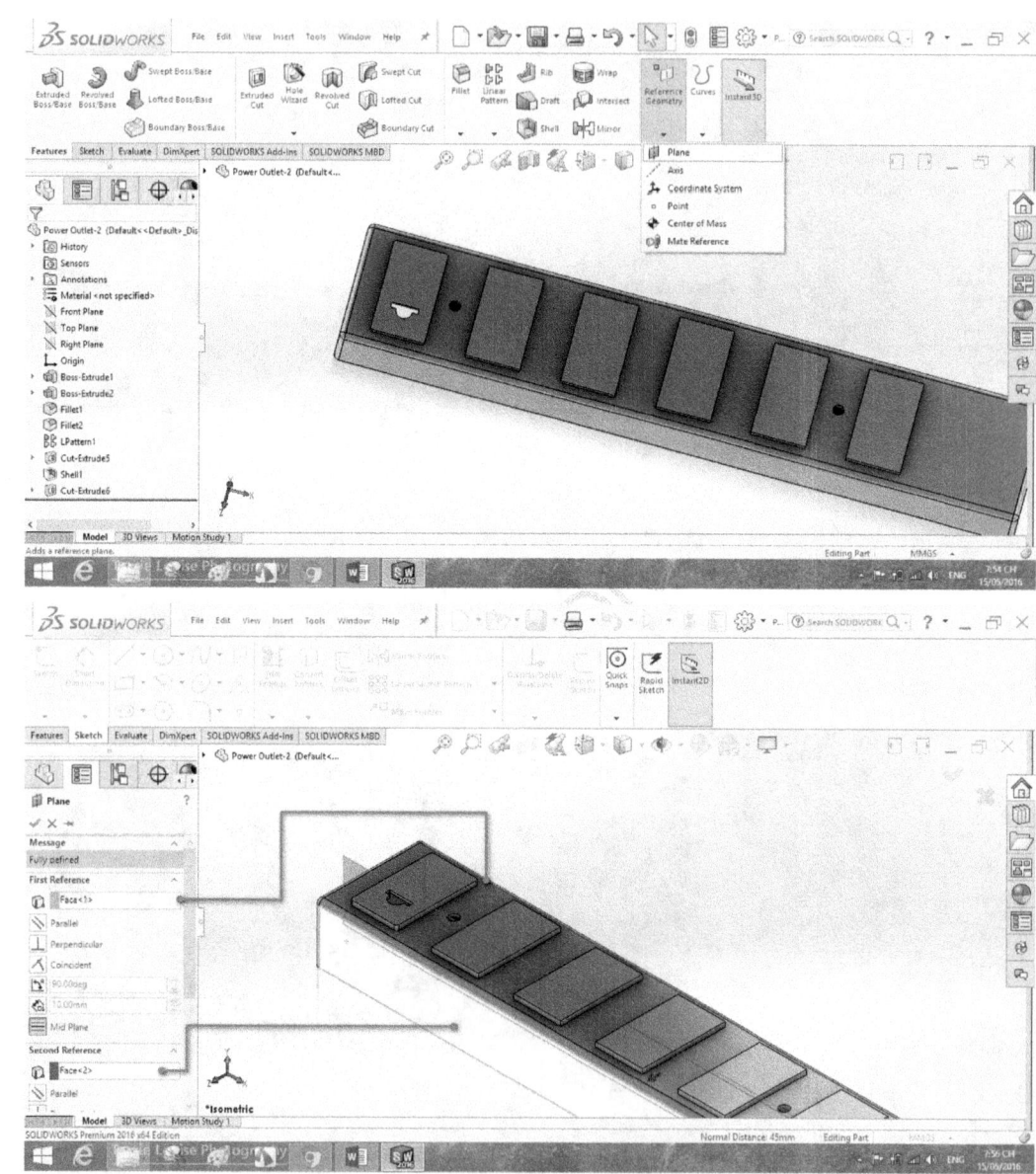

Step 12. Use Mirror. (Features=> Mirror). Then Select as below.

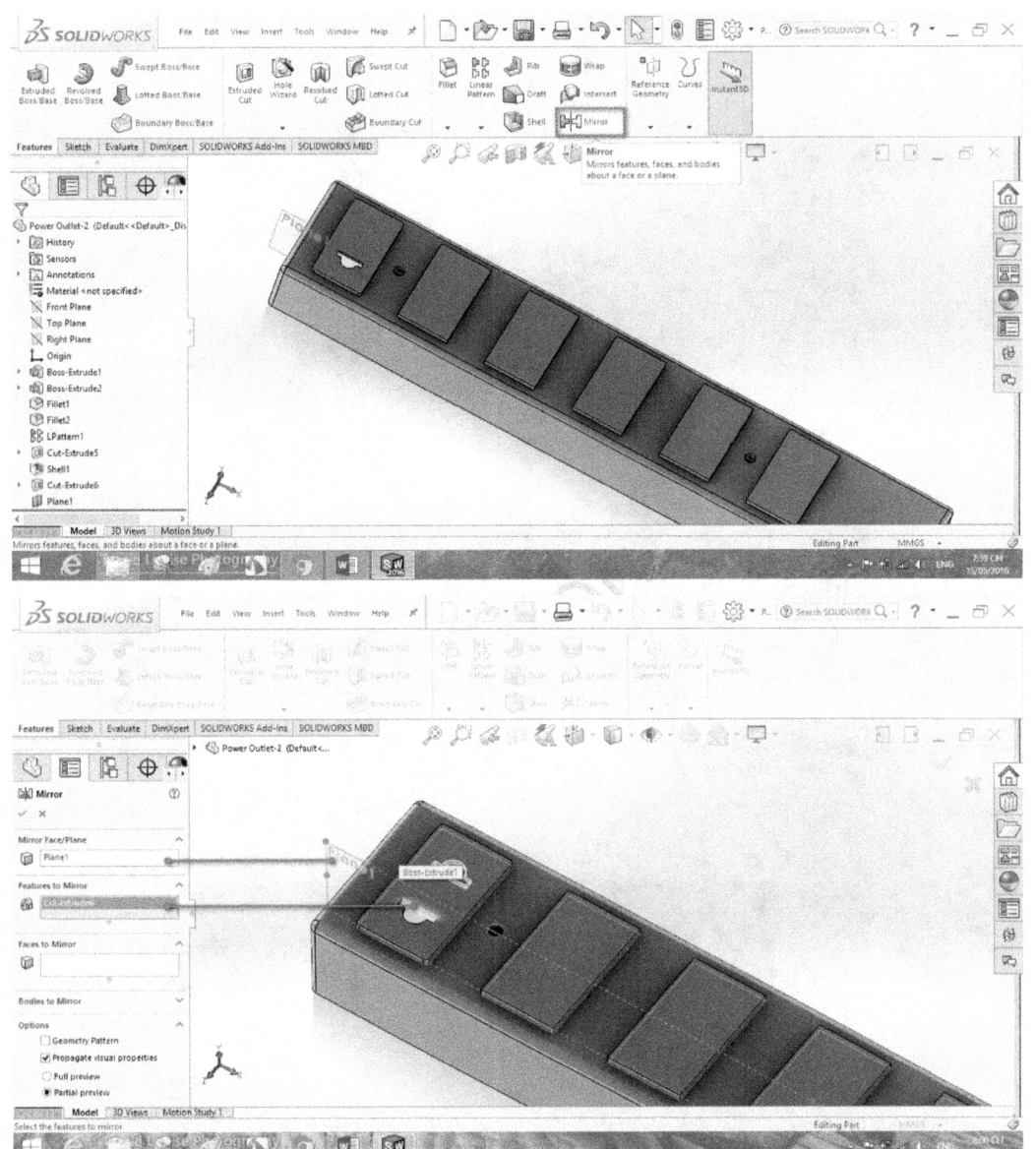

Step 13. Click Features=>Linear Pattern. Then Select as below.

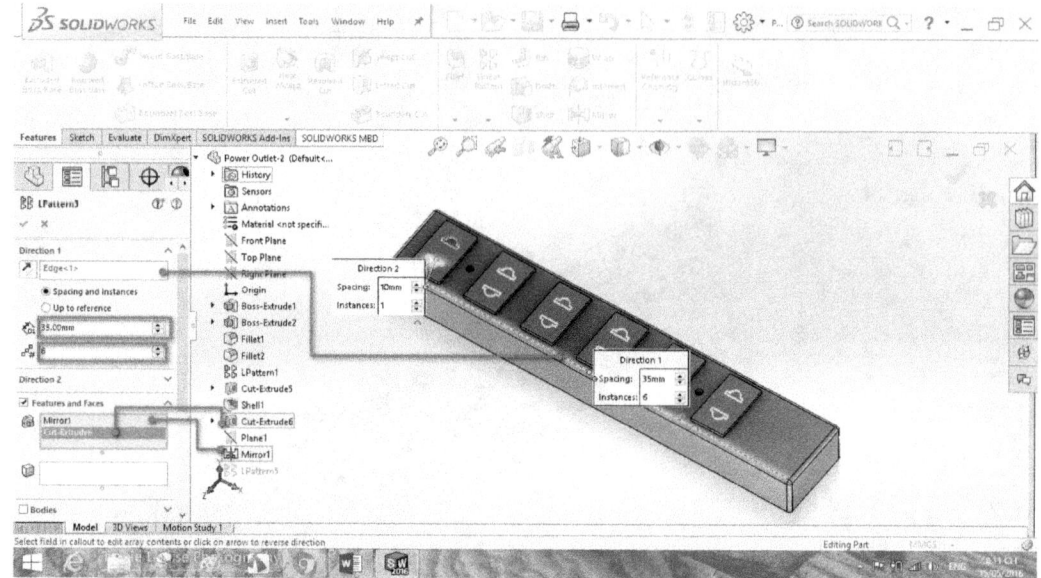

Step 14. Use Extruded Cut to Creat Cavity as below.

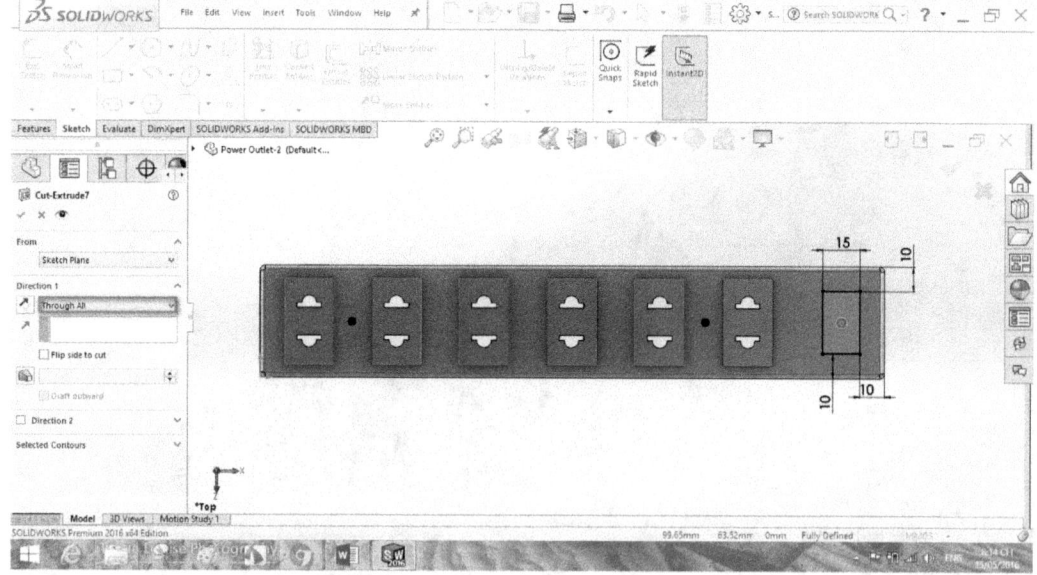

Step 15. Use Extruded Cut as below.

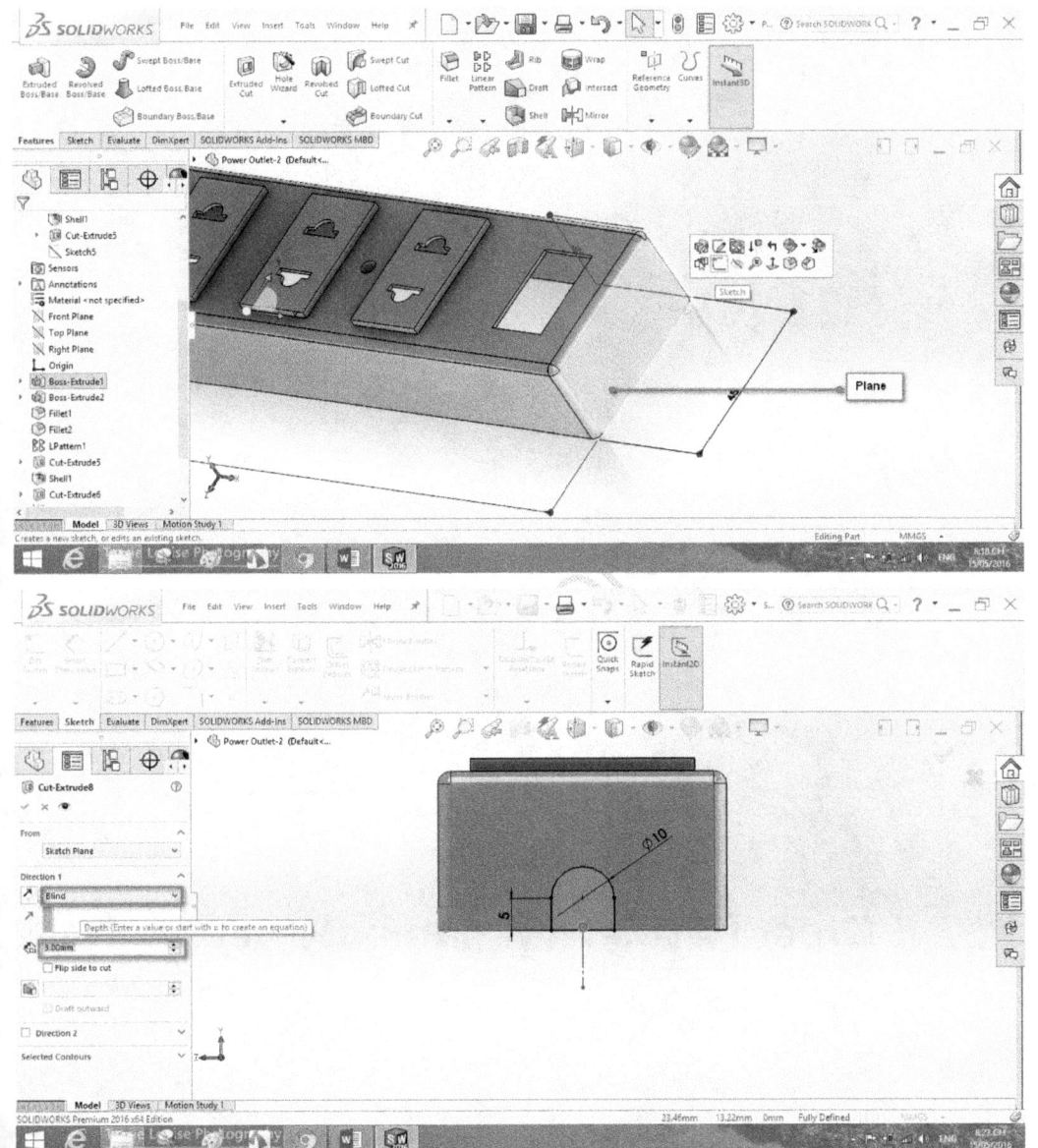

Step 16. Fillet the remaining edge. Save it with name Power Outlet-2.

Creat Part 3 with name Power Outlet-3

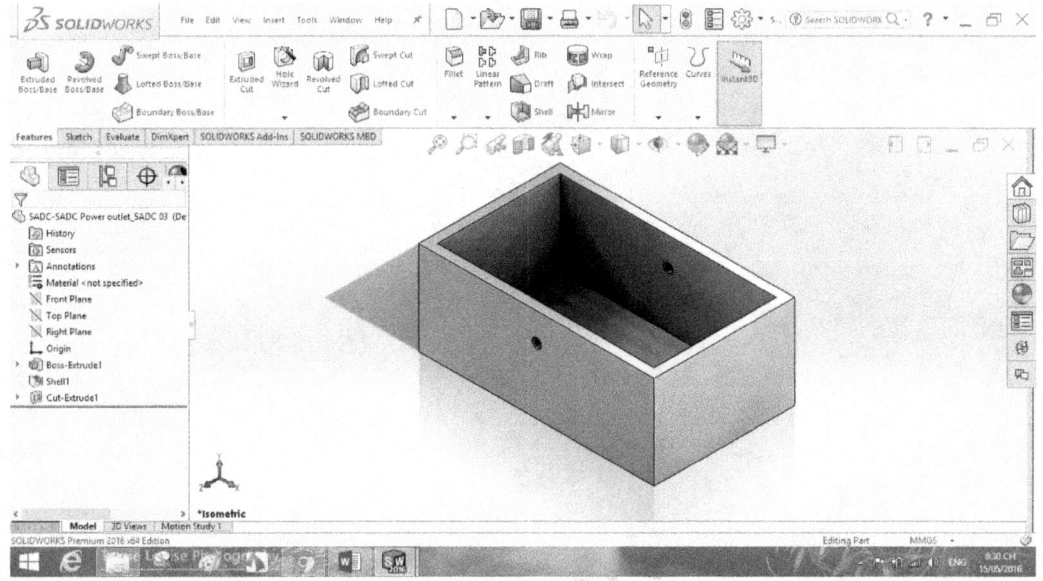

Step 1. Open Solidworks=>File=>New=>Part.

Step 2. Creat Sketch Rectangle (X=25, Y=15) On Top Plane, Use Extrude Boss with Depth=10.

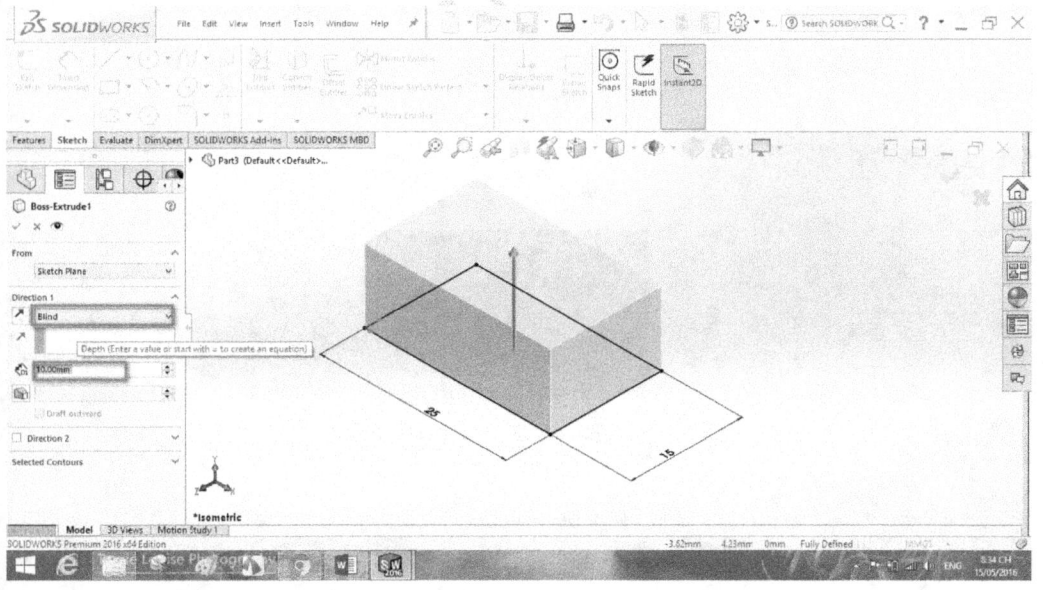

Step 3. Use Shell as below.

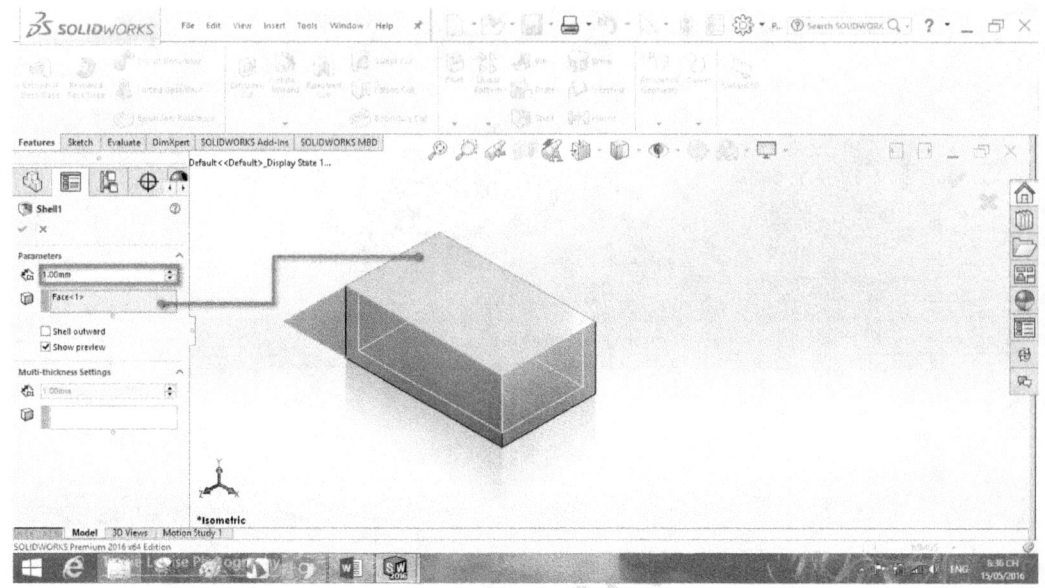

Step 4. Creat Sketch And Cut As Below. Save it with name Power Outlet-3

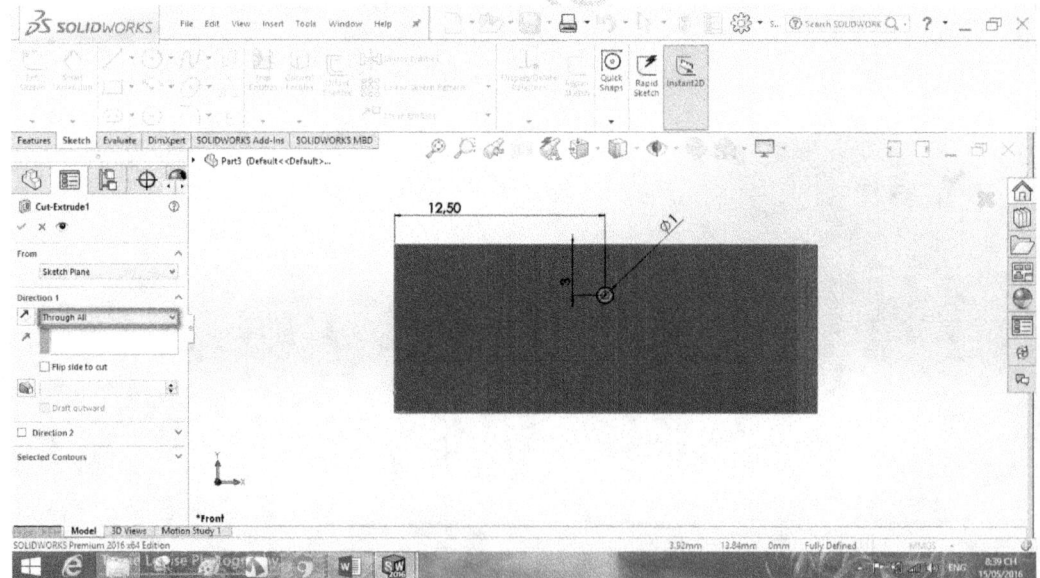

Creat Part 4 with name Power Outlet-4

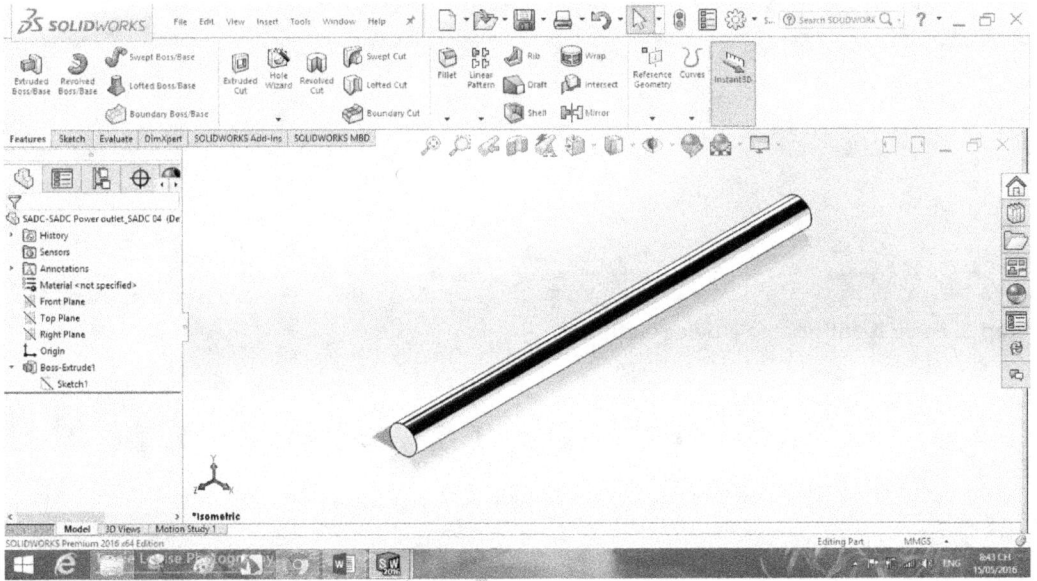

Step 1. Open Solidworks=>File=>New=>Part.

Step 2. Creat Sketch Circle (R=0.5) On Front Plane, Use Extrude Boss with Depth=15.

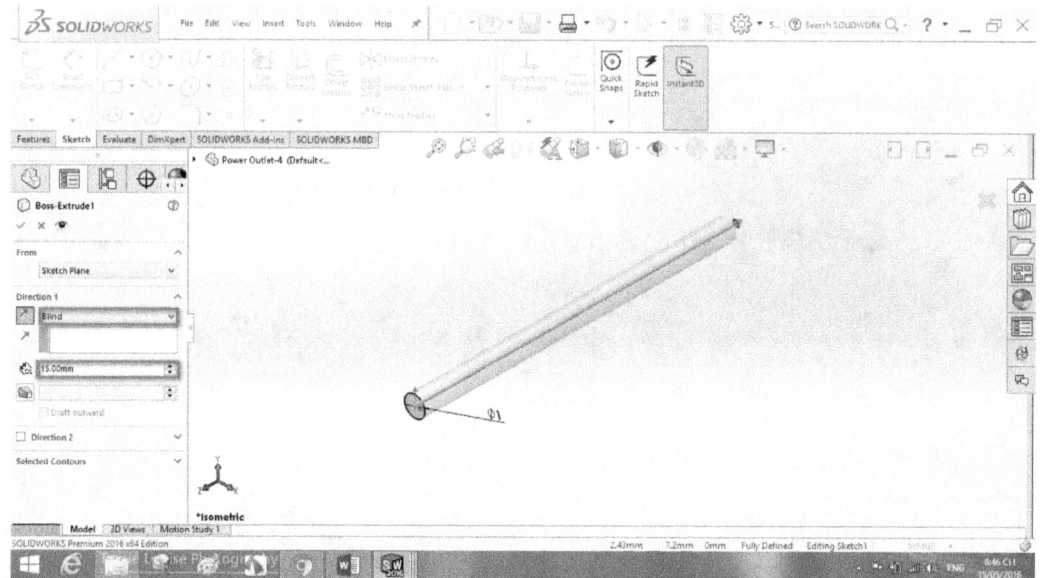

Step 3. Save it with name Power Outlet-4.

Creat Part 5 with name Power Outlet-5

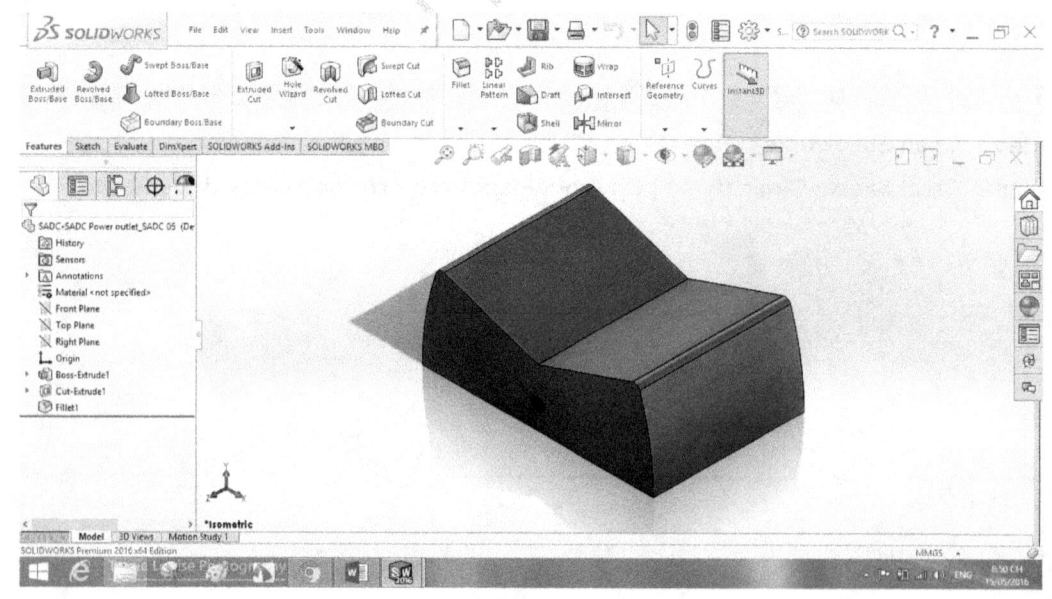

Step 1. Open Solidworks=>File=>New=>Part.

Step 2. Creat Sketch On Front Plane, Use Extrude Boss with Depth=13.

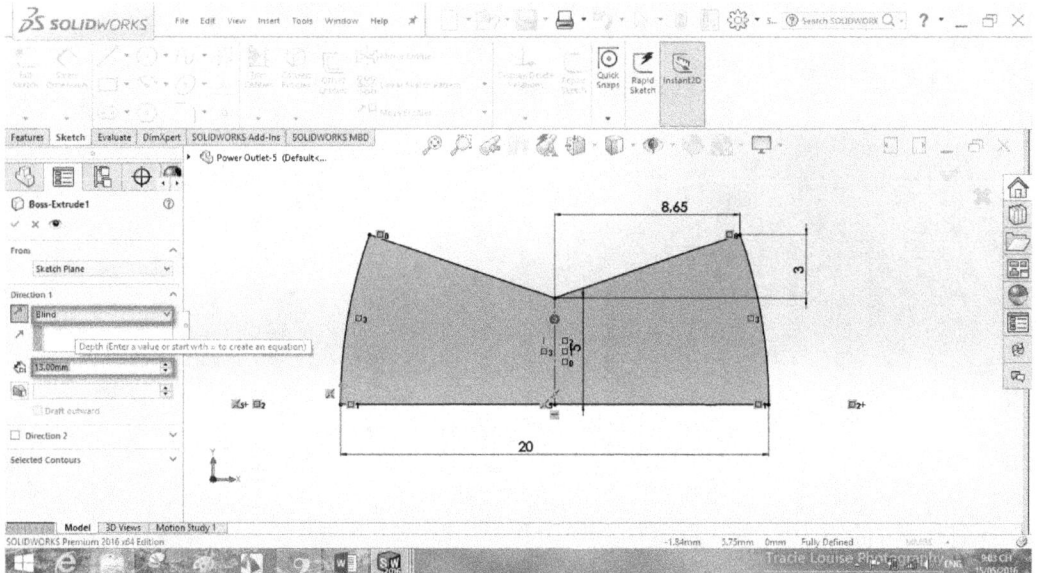

Step 3. Creat Hole Through All as below.

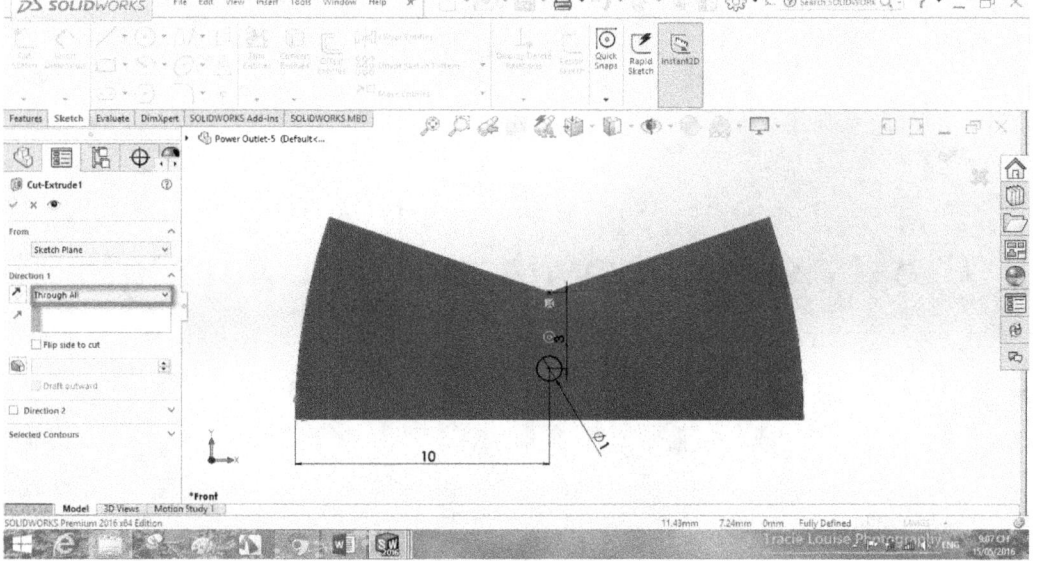

Step 5. Fillet R=0.5 as below. Save it with name Power Outlet-5.

Creat Assembly with name Power Outlet-5

Step 1. Open Solidworks=>File=>New=>Assembly.

Step 2. Click Browse=>Click Power Outlet-1=>Open then Click Ok.

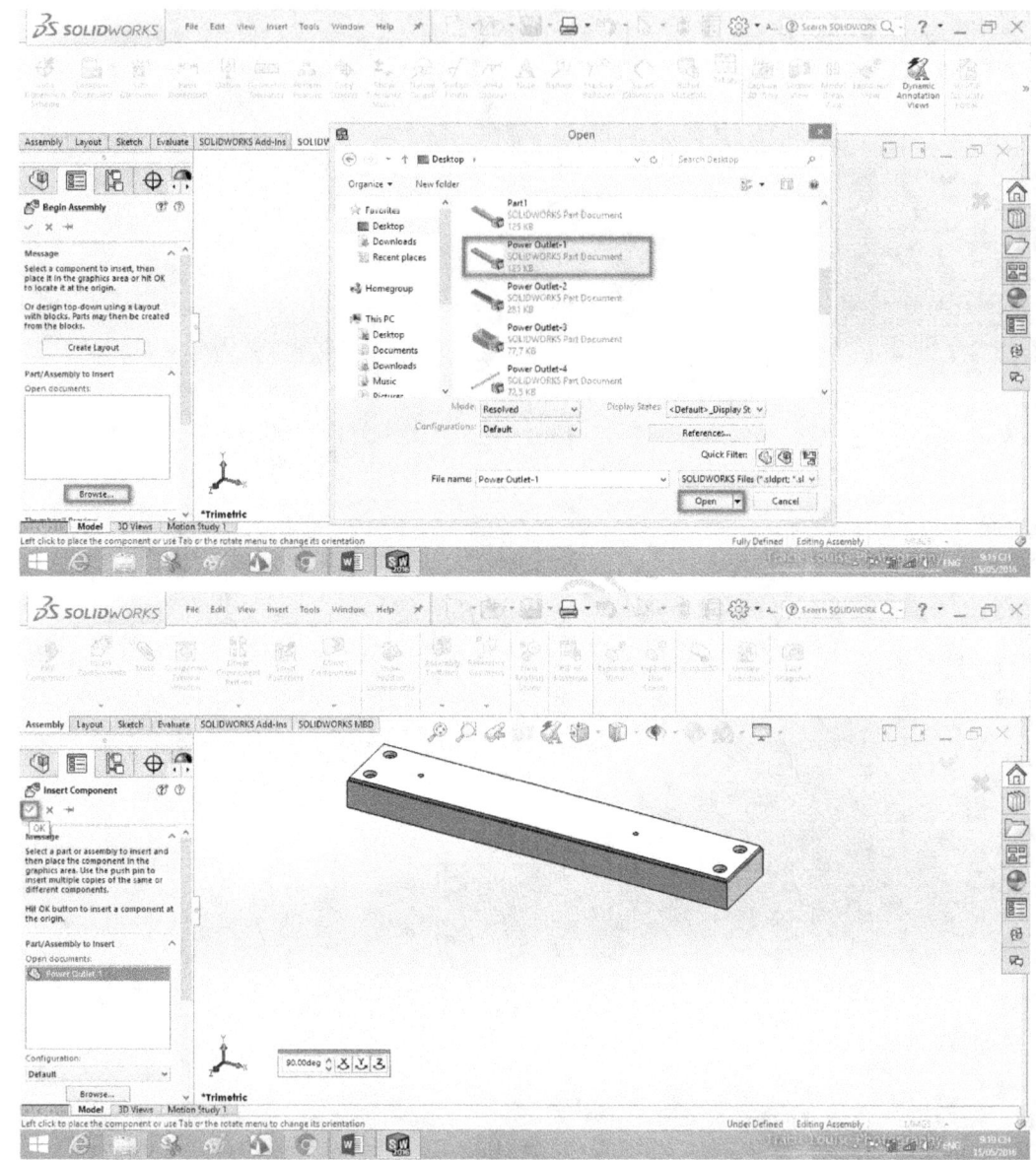

Step 3. Click Assembly=>Insert Components. Then select turn parts Power Outlet-2, Power Outlet-3, Power Outlet-4, Power Outlet-5. Put them outside the screen, No click OK. Because Power Outlet-1 to be fixed.

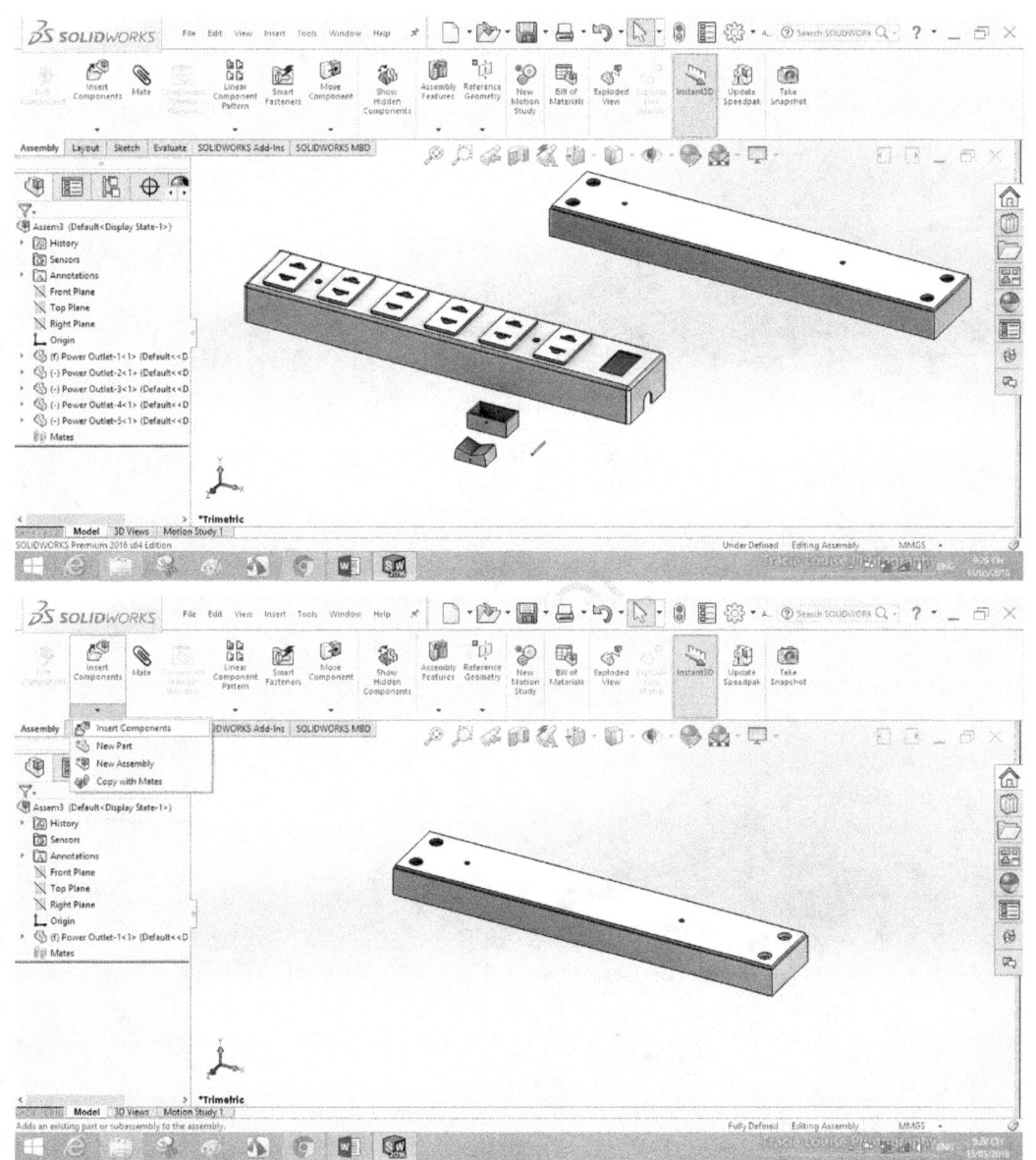

Step 4. Click Mate. Make a turn as below.

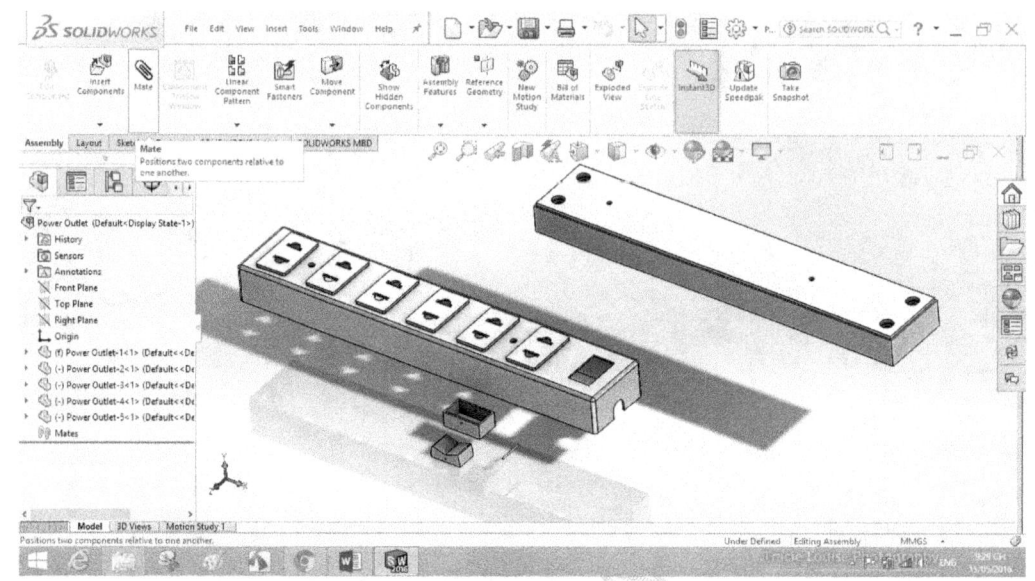

Step 5. Mate Power Outlet-1 with Power Outlet-2.

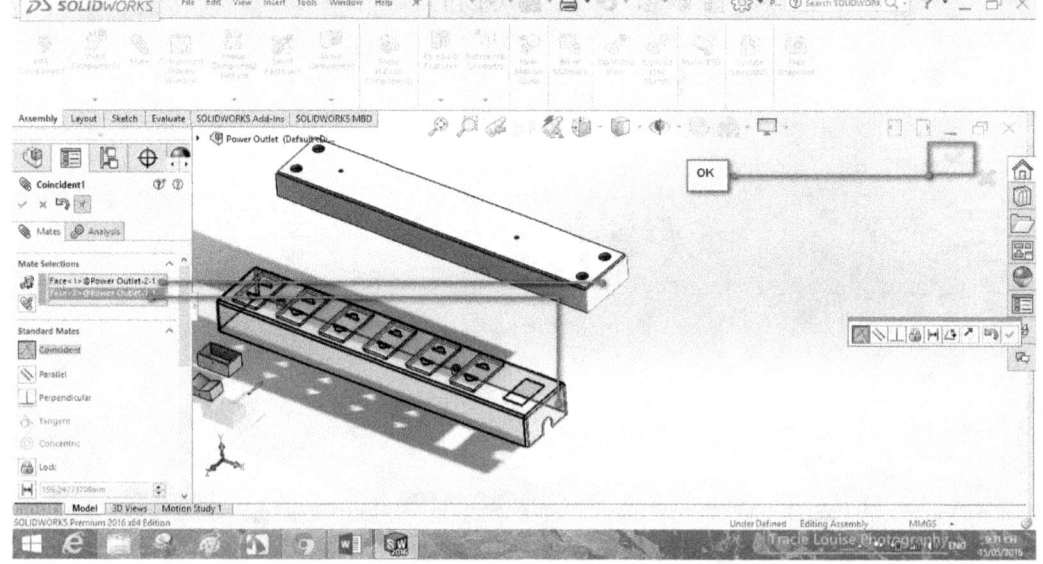

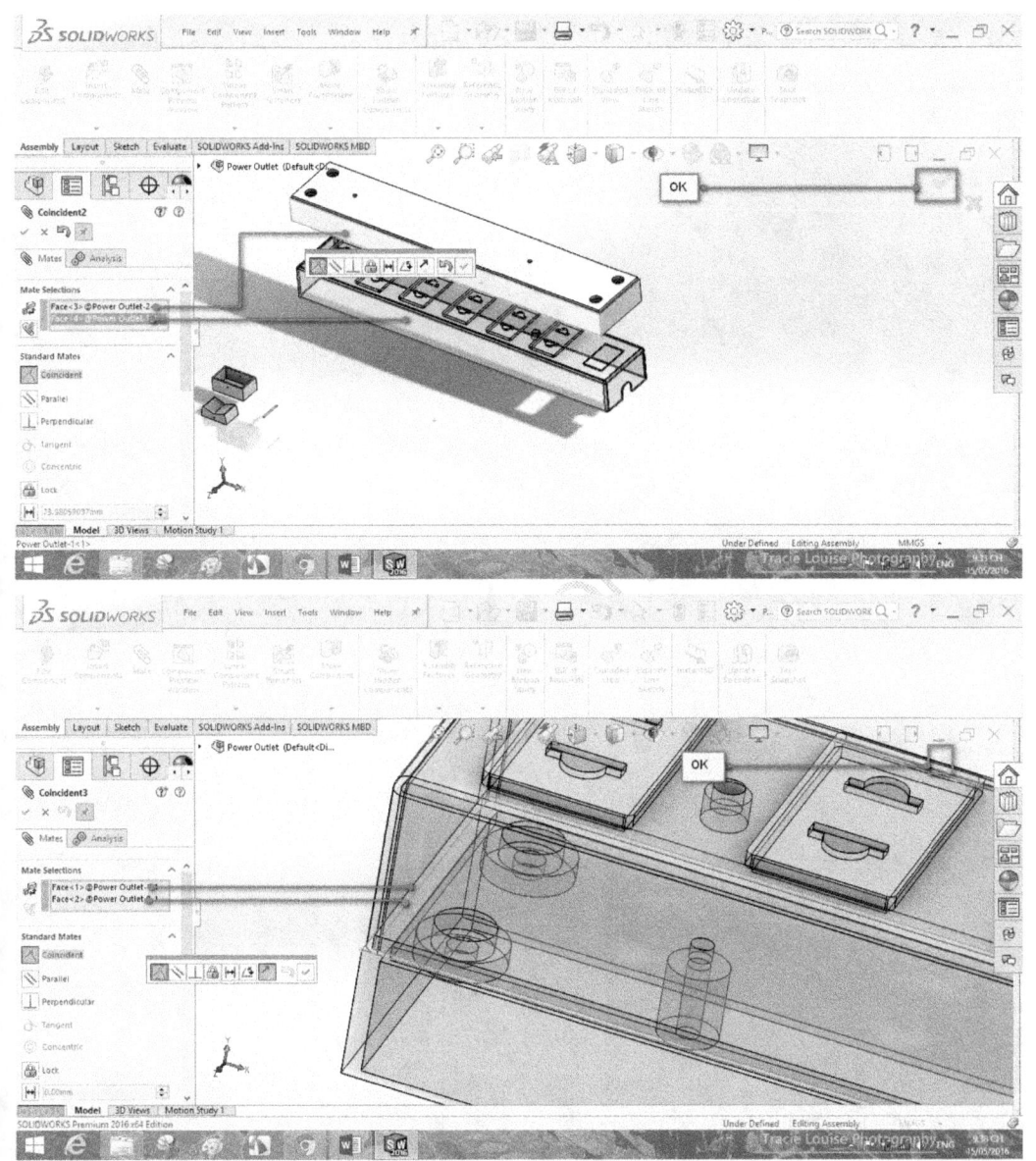

Step 5. Mate Power Outlet-2 with Power Outlet-3.

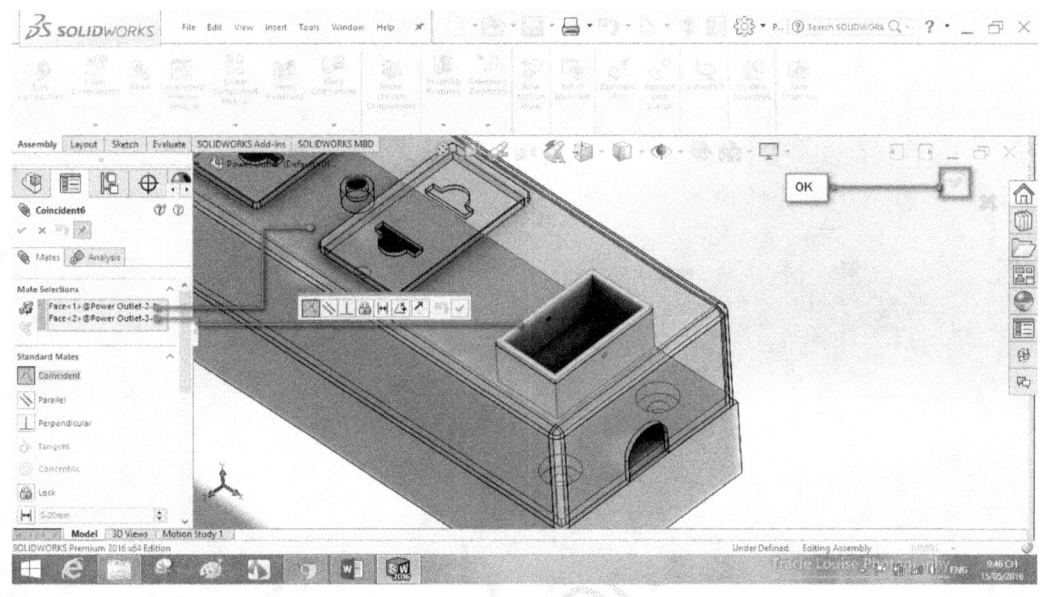

Step 5. Mate Power Outlet-3 with Power Outlet-4.

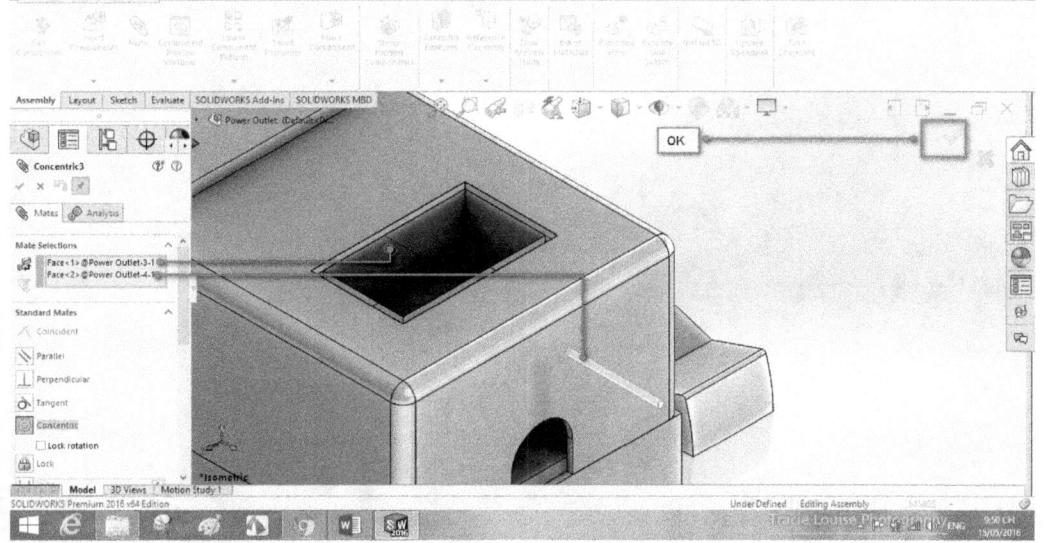

Step 6. Hide Components (Power Outlet-2)

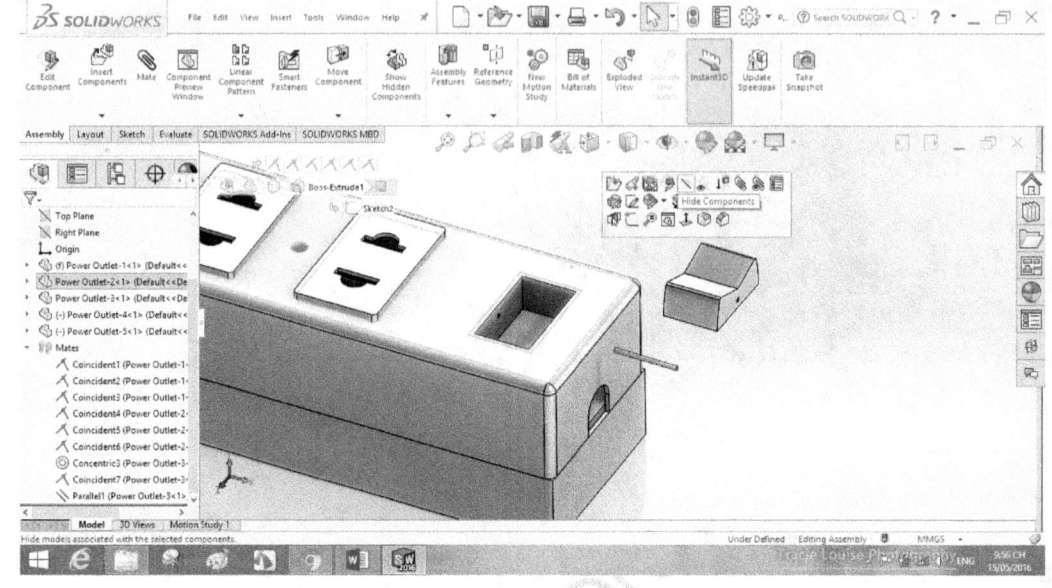

Step 7. Continue Mate Power Outlet-3 with Power Outlet-4.

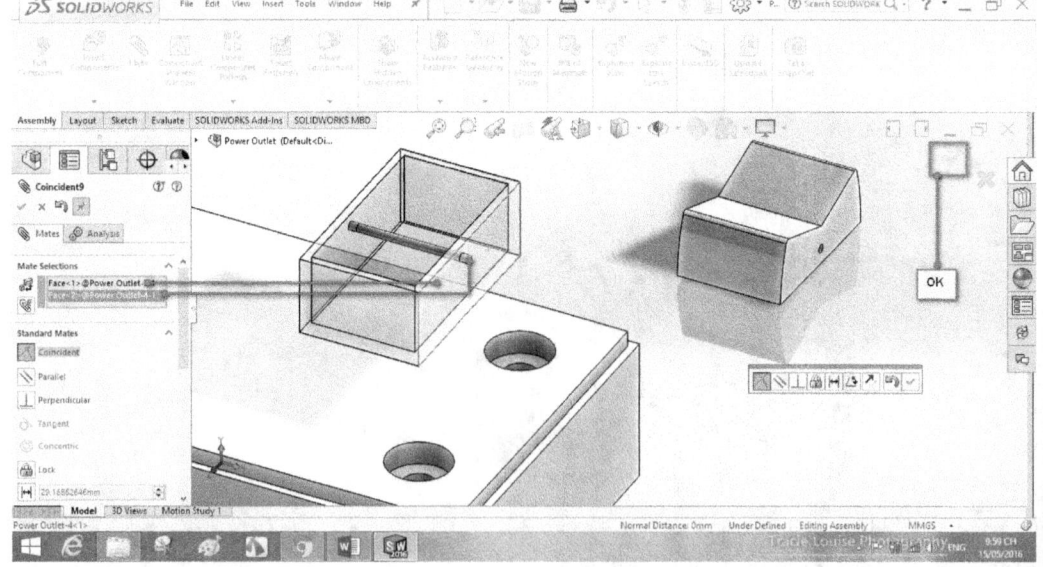

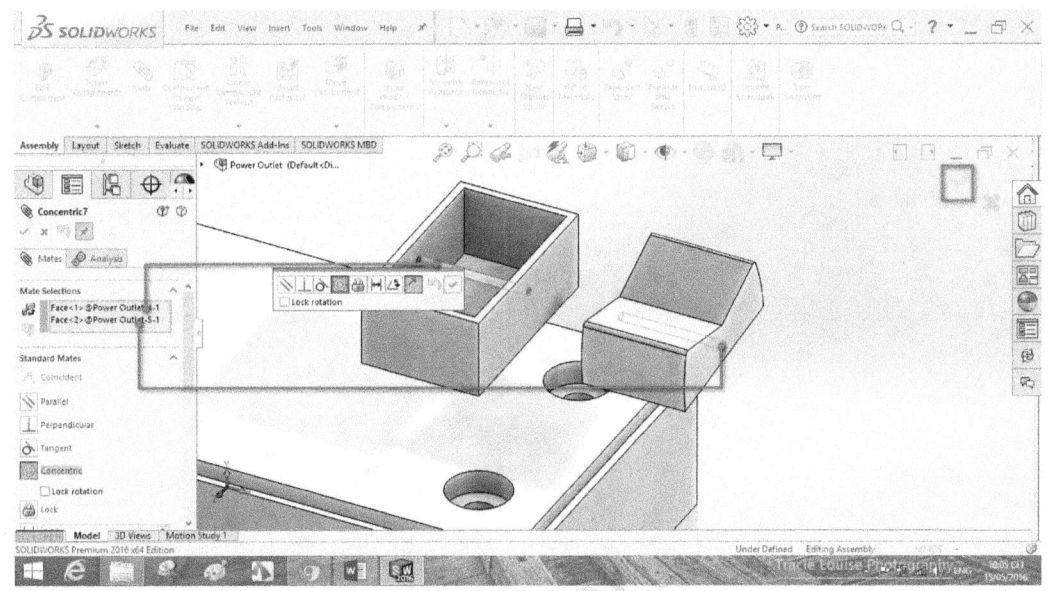

Step 8. Mate Power Outlet-3 with Power Outlet-5.

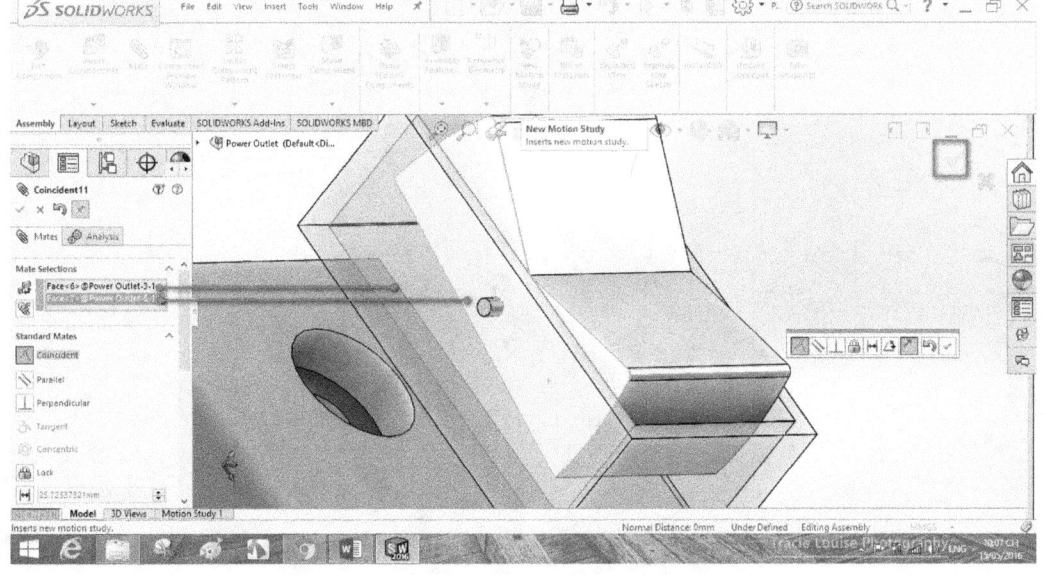

Step 9. Show Components (Power Outlet-2)

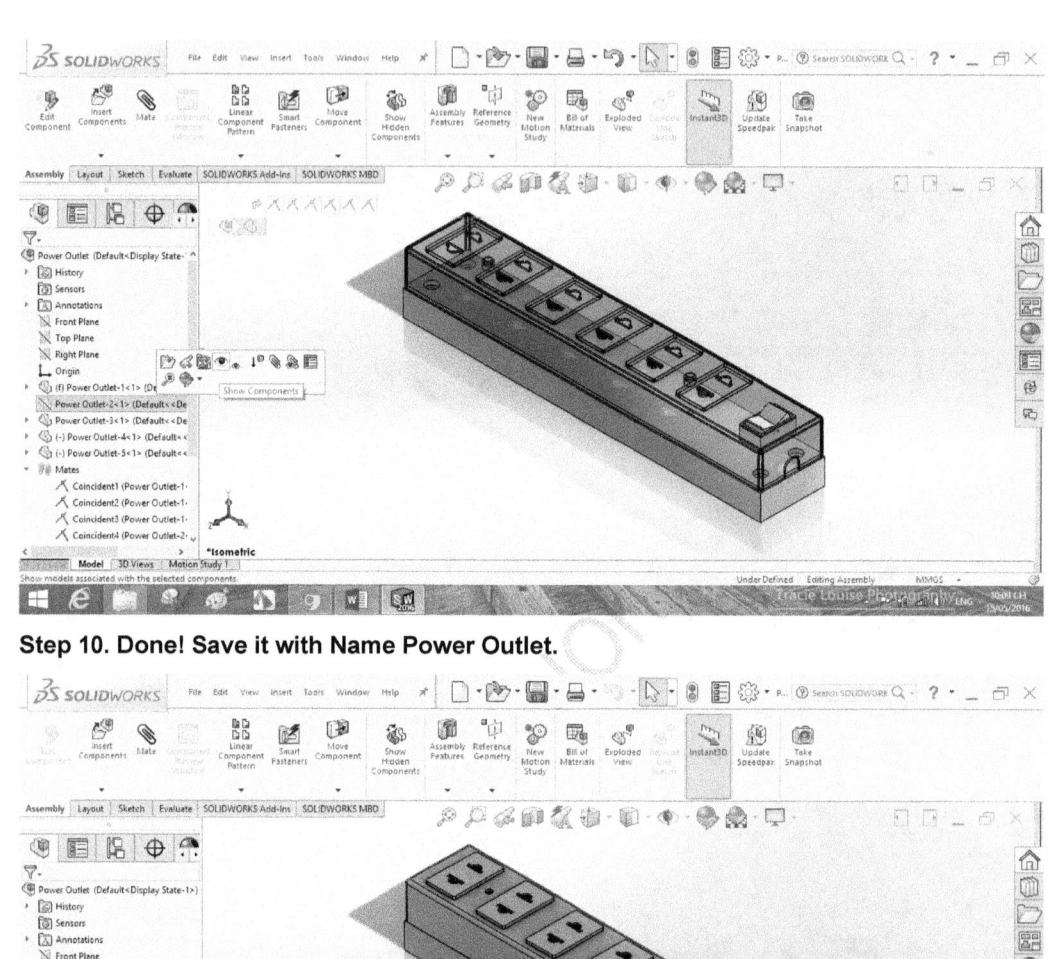

Step 10. Done! Save it with Name Power Outlet.

www.ingramcontent.com/pod-product-compliance
Lightning Source LLC
Chambersburg PA
CBHW070416190526
45169CB00003B/1285